FOUNDATIONS OF CHEMISTRY IN THE LABORATORY

THIRTEENTH EDITION

Morris Hein
Mount San Antonio College

Judith N. Peisen
Hagerstown Community College

Robert L. Miner
Mount San Antonio College

WILEY

John Wiley & Sons, Inc.

To order books or for customer service, please call 1-800-CALL-WILEY (225-5945).

ISBN-13   978-0-470-55490-6

Printed in the United States of America

10 9 8 7 6 5 4 3 2 1

Printed and bound by Bind-Rite, Inc.

Contents

EXPERIMENTS

STUDY AIDS

EXERCISES

APPENDICES

Preface

This manual is intended for the student who has not had a course in chemistry. The experiments are designed to be challenging but understandable to the student. Experimentation begins with simple laboratory techniques and measurements and progresses to relatively complex procedures. The 28 experiments are graded in difficulty to keep pace with the expanding capability of the student. The number and variety of experiments allow the instructor reasonable flexibility in preparing a laboratory schedule taht expands and supports many lecture topics in a one- or two-semester preparatory college chemistry course.

Our major objectives of this flexible laboratory program are to provide experience with (1) hands-on laboratory experimentation, (2) the capabilities and limitations of measurements, (3) a variety of chemical reactions and the equations used to describe them, (4) the collection, analysis, and graphing of data, (5) responsible disposal of chemicals for personal and environmental health, (6) Using a computer for graphing of data, (7) drawing valid conclusions form experimental evidence, and (8) support and reinforcement of concepts introduced in the lecture component of the course.

We have tried to maintain a balance between descriptive and quantitative experiments. All the experiments have been reviewed and modified to eliminate the use of heavy metals (where possible) and minimize exposure to hazardous material. Five experiments include unknowns for student analysis, and seven provide opportunities for graphing data. The Instructor's Manual provides sample student data, including graphs, for every experiment. The Instructor's Manual is available on the textbook website.

The format is designed to be helpful and convenient for both student and instructor and includes the following features:

1. A concise discussion of the basic underlying principles for each experiment provides pertinent background material to supplement, not replace, the textbook.

2. Six Study Aids provide supplementary material common to several experiments on the important topics of (a) significant figures, (b) formulas and chemical equations, (c) reading and preparing graphs by hand and by computer, (d) using a scientific calculator, (e) dimensional analysis and stoichiometry, and (f) introduction to organic chemistry.

3. Experimental procedures have been extensively tested by many students and provide enough detail for students to work with only general supervision.

4. Report forms for each experiment are cross-referenced to letters and subtitles in the procedure, designed to be completed before leaving the lab session, and relatively easy to grade.

5. The names and formulas of reagents used are listed at the beginning of each experiment.

6. Special safety precautions and waste disposal instructions are indicated when necessary at the point where they are required within the procedure.

7. For the convenience of the instructor and stockroom personnel, the appendices provide (a) an experiment-by-experiment list of special equipment and preparations needed, (b) a list of suggested equipment for student lockers, (c) an experiment-by-experiment list of waste disposal instructions, (d) a list of suggested auxiliary equipment, and (e) a complete list of reagents and details for the preparation of solutions.

The experiment which is new to the Thirteenth Edition is Experiment 8, Water, Solutions, and pH. This experiment provides an introduction to the properties of water especially important in the life sciences, introduces some skills used in biotechnology and the concept of molarity. This new experiment is an excellent foundation for Experiment 9, Properties of Solutions which is found in previous editions. Properties of Lead (II), Silver, and Mercury (I) Ions (Experiment 14 in the twelfth edition) which involved the use of heavy metal cations, has been eliminated.

Meticulous instructions for waste disposal have been continued and updated for students within each procedure and in the Instructor's Manual. The instructions for Preparing a Graph (Study Aid 3) have been updated to the most recent version of Excel (2007).

We are especially indebted to students in the chemistry departments of Mount San Antonio College and Hagerstown Community College for their patience and helpful suggestions during the development and testing of this laboratory program. We appreciate the feedback from instructors and students at the many schools over the years that have used this lab manual in their introductory chemistry course. A special thanks to Dr. Richard Montgomery, Dr. William Elliott, and Dr. Melanie Ulrich for their contribution to the Water, Solutions, and pH experiment that is new to this edition. Further suggestions for improvements of material in this laboratory manual are always welcome.

M. Hein
J. N. Peisen
R. L. Miner

To the Student

Since your laboratory time is limited, it is important to come to each session prepared by at least one hour of detailed study of the scheduled experiment. This should be considered a standing homework assignment.

Each of the experiments in this manual is composed of four parts:

1. **Materials and Equipment**—a list that includes the formulas of all substances used.

2. **Discussion**—a brief discussion of the principles underlying the experiment.

3. **Procedure**—detailed directions for performing the experiment with safety precautions clearly noted and disposal procedures for chemical waste provided throughout and identified by a waste icon. WASTE DISPOSE OF PROPERLY

4. **Report for Experiment**—a form for recording data and observations, performing calculations, and answering questions.

Follow the directions in the procedure carefully, and consult your instructor if you have any questions. For convenience, the letters and subtitles in the report form have been set up to correspond with those in the procedure section of each experiment.

As you make your observations and obtain your data, record them on the report form. Try to use your time efficiently; when a reaction or process is occurring that takes considerable time and requires little watching, start working on other parts of the experiment, perform calculations, answer questions on the report form, or clean up your equipment.

Except when your instructor directs otherwise, you should do all the work individually. You may profit by discussing experimental results with your classmates, but in the final analysis you must rely on your own judgment in completing the report form.

⚠ Safety Guidelines

While in the chemistry laboratory, you are responsible not only for your own safety but for the safety of everyone else. *We have included safety precautions in every experiment where needed, and they are highlighted with the icon shown in the title of this section.* Your instructor may modify these instructions and give you more specific directions on safety in your laboratory. If the proper precautions and techniques are used, none of the experiments in this laboratory program are hazardous. But without your reading and following the instructions, without knowledge about handling and disposal of chemicals, and without the use of common sense at all times, accidents can happen. Even when everyone is doing his or her best to comply with the safety guidelines in each experiment, accidents can happen. It is your responsibility to minimize these accidents and know what to do if they happen.

Laboratory Rules and Safety Procedures

1. **Wear protective goggles or glasses** at all times in the laboratory work area. These glasses should wrap around the face so liquids cannot splash into the eye from the side. These goggles are mandated by eye-protection laws and are not optional, even though they maybe uncomfortable. Contact lenses increase the risk of problems with eye safety, even when protective goggles are worn. If you wear contact lenses, inform the instructor.

2. **Dress appropriately** for the laboratory. Shoes that do not completely cover the foot are not allowed (*no sandals*). Long hair should be tied back. Wear a laboratory coat or apron, if available, to protect your clothing.

3. **Keep your benchtop organized as you work.** Put jackets, book bags, and personal belongings away from the work areas. Before you leave, clean your work area and make sure the gas and water are turned off. Clean and return all glassware and equipment to your drawer or the lab bench where you borrowed it.

4. **Keep all stock bottles of solid and liquid reagents in the dispensing area.** Do not bring reagent bottles to your laboratory work area. Use test tubes, beakers, or weigh boats to obtain chemicals from the dispensing areas: (1) the reagent shelf, (2) the balance tables, (3) under the fume hood, and (4) as instructed.

5. **Keep the balance and the area around it clean.** Do not place chemicals directly on the balance pans; place a piece of weighing paper, a weigh boat, or another small container on the pan first, and then weigh your material. **Never weigh an object while it is hot.**

6. **Carefully check the name on the reagent bottles before you use them.** Many names and formulas appear similar at first glance. Label every beaker, test tube, etc., into which you transfer chemicals. Many labels will contain the National Fire Protection Association (NFPA) diamond label, which provides information about the flammability, reactivity, health effects, and miscellaneous effects for the substance. Each hazard is rated 0 (least hazardous) to 4 (most hazardous). For example, the NFPA label for potassium chromate is shown below.

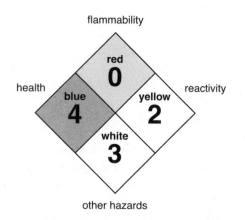

More specific information (the reason for potassium chromate being rated an extreme health hazard, for example) about all known substances is available in the form of Material Safety Data Sheets (MSDS), which many institutions keep on file for chemicals stored and used in their laboratories. MSDS are usually provided with chemicals by the supplier when they are purchased and are easily obtained from many website sources. Because of its hazardous nature, chromates have been removed from this lab manual.

7. **Never return unused chemicals to the reagent bottles.** This is a source of possible contamination of the entire stock bottle. Dispose of unused chemicals exactly as instructed in the waste disposal instructions for that substance, identified by [WASTE DISPOSE OF PROPERLY] throughout each experiment.

8. **Disposal of wastes must follow state and federal guidelines.** Do not put anything into the trash or sink without thinking first. We have tried to anticipate every disposal decision in the procedure and marked the procedure with the waste icon. The following guidelines are the foundation of waste disposal decisions:

    a. Broken glass is put into a clearly marked special container.

    b. Organic solvents are never poured into the sink. They are usually flammable and often immiscible with water. Instead, they are poured into a specially marked container ("waste organic solvents") provided when needed.

    c. Solutions containing cations and anions considered toxic by the EPA are never poured into the sink. They are poured into specially marked containers ("waste heavy metal," etc.) provided when needed. The name of all ions disposed of into a specific bottle must be listed on the label.

    d. Solutions poured in the sink should be washed down with plenty of water.

    e. Some solid chemicals must also be disposed of in specially labeled containers. If you are not sure what to do, ask the instructor.

    f. Each school may have its own policy for waste disposal which supercedes the instructions in this manual.

9. **Avoid contaminating stock solutions.** Do not insert medicine droppers or pipets into reagent bottles containing liquids. Instead, pour a little of the liquid into a small beaker or test tube. If the bottle is fitted with a special pipet that is stored with the bottle, this may not be necessary.

10. **Avoid all direct contact with chemicals.**

    a. Wash your hands anytime you get chemicals on them and at the end of the laboratory session.

    b. If you spill something, clean it up immediately before it dries or gets on your papers or skin.

    c. Never pipet by mouth.

    d. Never eat, drink, or smoke in the laboratory.

    e. Do not look down into the open end of a test tube in which a reaction is being conducted, and do not point the open end of a test tube at someone else.

    f. Inhale odors and chemicals with great caution. Waft vapors toward your nose. The fume hood will be used for all irritating and toxic vapors.

11. **Working with glass requires special precautions:**

    a. Do not heat graduated cylinders, burets, pipets, or bottles with a burner flame.

    b. Do not hold a test tube or beaker in your hand during a chemical reaction.

c. Do not touch glass that has been near a flame or hot plate. Hot glass looks the same as cool glass and may cause serious burns.

d. Learn and practice proper procedures when inserting glass tubing into rubber stoppers.

12. **Learn the location and proper use of safety equipment:** fire extinguisher, eye wash, first aid kit, fire blanket, safety shower, spill kits, and other equipment available.

13. **Never work alone** in the laboratory area.

14. **Report all accidents** to the instructor, no matter how minor.

15. **Do not perform unauthorized experiments.**

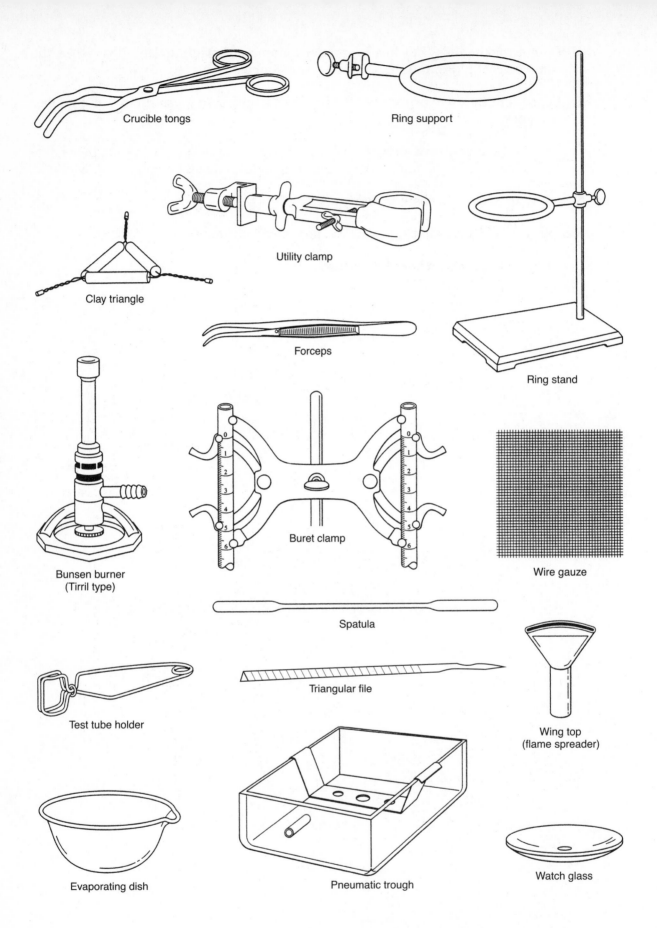

Crucible tongs

Ring support

Utility clamp

Clay triangle

Forceps

Ring stand

Bunsen burner
(Tirril type)

Buret clamp

Wire gauze

Spatula

Test tube holder

Triangular file

Wing top
(flame spreader)

Evaporating dish

Pneumatic trough

Watch glass

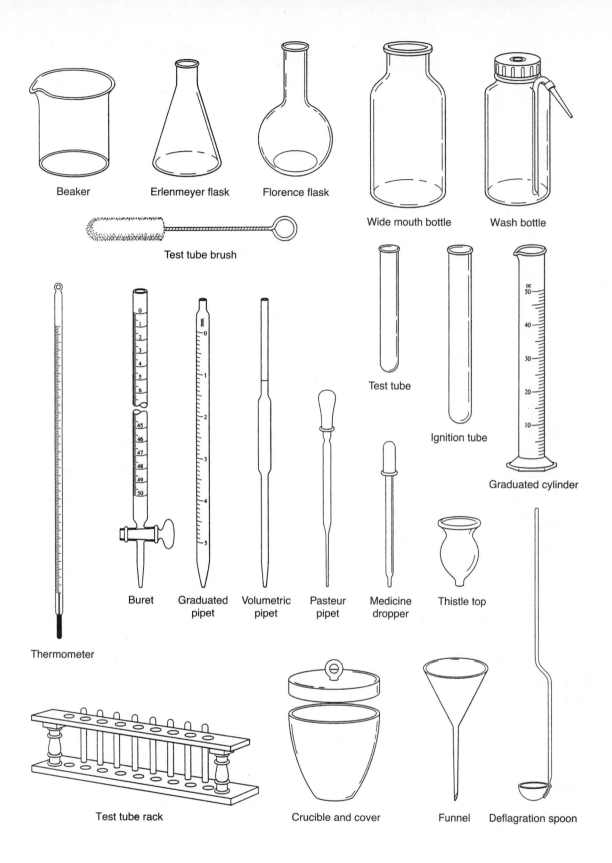

Beaker

Erlenmeyer flask

Florence flask

Wide mouth bottle

Wash bottle

Test tube brush

Test tube

Ignition tube

Graduated cylinder

Thermometer

Buret

Graduated pipet

Volumetric pipet

Pasteur pipet

Medicine dropper

Thistle top

Test tube rack

Crucible and cover

Funnel

Deflagration spoon

EXPERIMENT 1

Laboratory Techniques

MATERIALS AND EQUIPMENT

Solids: lead(II) iodide (PbI_2), sodium nitrate ($NaNO_3$), and sodium chloride ($NaCl$). **Liquid:** glycerol. **Solutions:** 0.1 M lead(II) nitrate [$Pb(NO_3)_2$] and 0.1 M sodium iodide (NaI). Ceramfab pad, 100 mL and 400 mL beakers, Bunsen burner, No. 1 evaporating dish, triangular file, funnel, wire gauze, filter paper, glass rod, clay triangle, 6 mm glass tubing, wing top (flame spreader).

DISCUSSION AND PROCEDURE

Wear protective glasses.

A. Laboratory Burners

Almost all laboratory burners used today are modifications of a design by the German chemist Robert Bunsen. In Bunsen's fundamental design, also widely used in domestic and industrial gas burners, gas and air are premixed by admitting the gas at relatively high velocity from a jet in the base of the burner. This rapidly moving stream of gas causes air to be drawn into the barrel from side ports and to mix with the gas before entering the combustion zone at the top of the burner.

The burner is connected to a gas cock by a short length of rubber or plastic tubing. With some burners the gas cock is turned to the **fully on** position when the burner is in use, and the amount of gas admitted to the burner is controlled by adjusting a needle valve in the base of the burner. In burners that do not have this needle valve, the gas flow is regulated by partly opening or closing the gas cock. With either type of burner **the gas should always be turned off at the gas cock when the burner is not in use** (to avoid possible dangerous gas leakage from the needle valve or old tubing).

1. **Operation of the Burner.** Examine the construction of your burner (Figure 1.1) and familiarize yourself with its operation. A burner is usually lighted with the air inlet ports nearly closed. The ports are closed by rotating the barrel of the burner in a clockwise direction. After the gas has been turned on and lighted, the size and quality of the flame is adjusted by admitting air and regulating the flow of gas. Air is admitted by rotating the barrel; gas is regulated with the needle valve, if present, or the gas cock. Insufficient air will cause a luminous yellow, smoky flame; too much air will cause the flame to be noisy and possibly blow out. A Bunsen burner flame that is satisfactory for most purposes is shown in Figure 1.2; such a flame is said to be "nonluminous." Note that the hottest region is immediately above the bright blue cone of a well-adjusted flame.

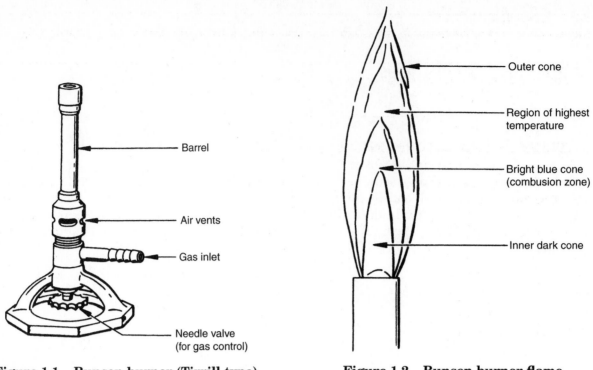

Figure 1.1   Bunsen burner (Tirrill type)

Barrel

Air vents

Gas inlet

Needle valve
(for gas control)

Figure 1.2   Bunsen burner flame

Outer cone

Region of highest
temperature

Bright blue cone
(combusion zone)

Inner dark cone

B.   Glassworking

WASTE DISPOSE OF PROPERLY   **Dispose of broken and nonusable glass in the container provided.**

In laboratory work it is often necessary to fabricate simple items of equipment, making use of glass tubing and rubber stoppers. In working with glass tubing, improper techniques may result not only in an unsatisfactory apparatus but also in severe cuts and burns. Therefore the numbered instructions below should be studied carefully. Prepare the following list of items (illustrated in Figure 1.3), using 6 millimeter (mm) glass tubing and rod.

Two straight tubes, one 24 centimeters (cm) long, the other 12 cm (Figure 1.3A).

Two right-angle bends (Figure 1.3B).

One delivery tube (Figure 1.3C).

Two buret tips (Figure 1.3D). (Optional)

One stirring rod if there is none in your locker (Figure 1.3E).

This equipment will be used in future experiments. After it has been completed and approved by your instructor, store them in your locker.

1. **Cutting Glass Tubing.**   (See Figure 1.4.) Mark the tube with a pencil or ball-point pen at the point where it is to be cut. Grasp the tubing about 1 cm from the mark and hold it in position on the laboratory table. Hold the file by the tang (or handle) end and, pressing the edge of the file firmly against the glass at right angles to the tubing, make a scratch on the tubing by pushing the file away from you. If the file is in good condition a single stroke should suffice. Several strokes may be required if the file is dull, but if more than one stroke is need-ed, all must follow the same path so that only one scratch mark is present on the tubing. The scratch need not be very deep or very long, but it should be clearly defined.

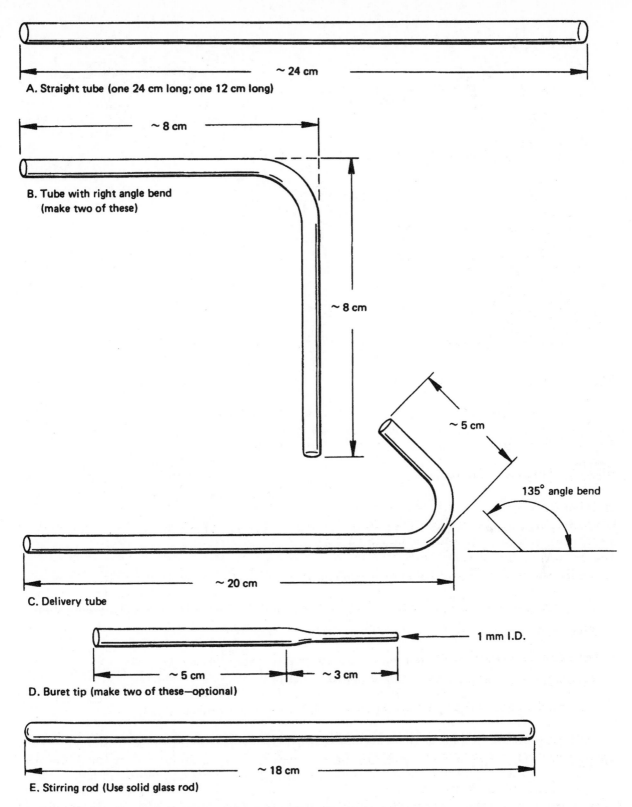

A. Straight tube (one 24 cm long; one 12 cm long)

~ 24 cm

~ 8 cm

B. Tube with right angle bend
(make two of these)

~ 8 cm

~ 5 cm

135° angle bend

~ 20 cm

C. Delivery tube

1 mm I.D.

~ 5 cm

~ 3 cm

D. Buret tip (make two of these—optional)

~ 18 cm

E. Stirring rod (Use solid glass rod)

Figure 1.3   Glassware (Illustrations are not to scale)

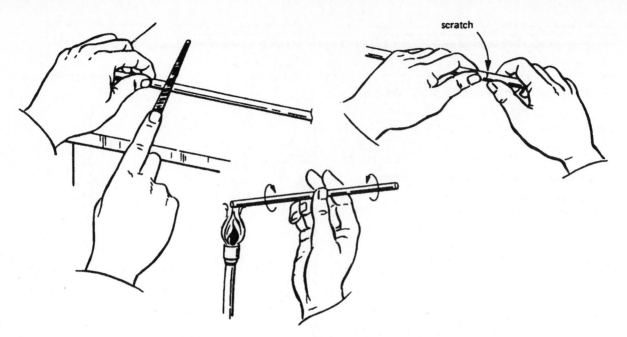

Figure 1.4   Cutting and fire-polishing glass tubing

Grasp the tubing with your thumbs together directly opposite the scratch mark (see Figure 1.4). Now apply pressure with the thumbs as though bending the ends toward your body while at the same time exerting a slight pull on the tubing. A straight, clean break should result. Use the flat side of your file to remove any sharp projections from the ends of the cut tubing. After cutting glass in this way, the ends of the cut glass, although clean and flat, are still very sharp and must be fire-polished in order to avoid personal injury.

2. **Fire-Polishing Glass.   Fire polishing** is the process of removing the sharp edges of glass by heating the tubing in a burner flame.

While continuously rotating the tubing, heat the end in the hottest part of the flame until the sharp edges are smooth. Be careful not to heat too much because the opening will become constricted. When the fire-polishing is completed, remember that the glass is **hot** even though it looks cool.

Put the hot glass tubing on a Ceramfab pad to cool. This is an excellent safety device. If hot objects are always placed on the pad and allowed to cool, then picked up with caution, one is less likely to get burned. The Ceramfab pad also protects the hot glass from sudden chilling (thermal shock) and the table top from injury.

Your instructor will have some examples of properly fire-polished tubing available for your inspection. Laboratory stirring rods are easily made by cutting glass rod in the same way described for tubing and fire-polishing the ends until they are smooth and rounded.

Whenever glass is cut it must be fire-polished in order to avoid personal injury.

3. **Bending Glass Tubing.**   Put the wing top (flame spreader) on your burner and adjust the flame so that a sharply defined region of intense blue color is visible. Grasp the tubing to be bent at both ends and hold it in the flame lengthwise just above the zone of intense blue color. Continuously rotate the tubing in the flame until it has softened enough to bend easily

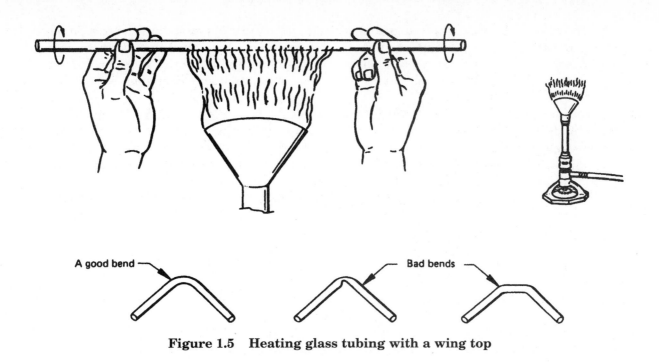

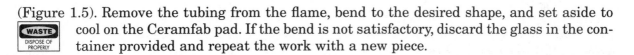

Figure 1.5   Heating glass tubing with a wing top

(Figure 1.5). Remove the tubing from the flame, bend to the desired shape, and set aside to cool on the Ceramfab pad. If the bend is not satisfactory, discard the glass in the container provided and repeat the work with a new piece.

If a bend is to be made where one arm of tubing is too short to hold in the hand while heating (Figure 1.3C), follow one of two procedures: (1) Proceed as in the above paragraph, using enough tubing to handle it from both ends; then cut to size and firepolish it after the bend is completed. (2) Heat a piece of tubing of proper size, holding it at one end and rotating it until it is soft in the region to be bent; then remove it from the flame and bend by grasping with tongs or by inserting the tang of a file into the hot end of the tubing.

4. **Preparing Buret Tips (Jets).**   A buret tip (or jet) (Figure 1.3D) is prepared as follows: Remove the wing top and heat a small section in the center of a 14 cm length of tubing while rotating it in the hottest portion of the burner flame (without wing top) until the tubing is very soft. Remove from the flame and slowly pull the ends away from each other while holding the tubing in a vertical position. After cooling, cut the tips to the desired dimensions. Fire-polish all edges.

5. **Inserting Glass Tubing into a Stopper.**

⚠ If the glass tubing is held in the wrong place during its insertion into a stopper, serious injury can occur if the tubing shatters or breaks and pushes into the hand. Read the following instructions carefully.

Lubricate the hole in the stopper with glycerol (glycerine), using a stirring rod to make sure that the lubricant actually gets into the hole. Grasp the fire-polished tubing about 1 cm from the end to be inserted. Holding the lubricated stopper in the other hand, start the tubing into the hole by gently twisting it, and gradually work it all the way through the stopper. Be sure to grip the tubing at a point not more than 1 cm from the stopper at all times when making the insertion. Gripping at greater distances and twisting will break the tubing and probably cause personal injury. It is also good safety practice to protect your hands with a towel when inserting or removing glass tubing from rubber stoppers.

The end of the tube should protrude at least 5 mm from the stopper, so that the free passage of fluids is not prevented by flaps of rubber. After insertion make sure that the tube is not plugged, and remove the excess glycerol either by washing or with a towel. Note the lubricating properties of glycerol by rubbing a drop between your thumb and forefinger.

3. **Removing Glass Tubing from Stoppers.** Tubing should be removed from stoppers at the end of the laboratory period. Be careful to grip the tubing at a point very close to the stopper when twisting. If allowed to stand for several days the stopper may stick to the glass and be difficult to remove. In the event of sticking, do not use "strong arm" methods. A cork borer (No. 3 for 6 mm tubing) may be inserted between the rubber and glass to help remove the tubing from the stopper, but consult your instructor before attempting this procedure.

C. Evaporation

Evaporation is one of the processes used to separate a dissolved solid from a liquid.

1. Prepare the simple water bath illustrated in Figure 1.6. Before placing the beaker into position be sure the hottest part of the burner flame will reach the bottom of the beaker. Put the beaker of tap water on the wire gauze and begin heating.

2. **Preparing the Solution.** Cover the bottom of a 150 mm (standard) test tube with a small quantity of sodium chloride. Add distilled water until the test tube is about one-quarter full and stir with a glass rod until the salt is dissolved.

3. **Evaporating the Solution.** Pour the sodium chloride solution into an evaporating dish. Place the evaporating dish on the beaker of water being heated. Continue heating the water to maintain boiling—replenishing if necessary—until all of the water has evaporated from the solution in the evaporating dish, leaving the original solid as the residue.

4. Dissolve the residue with tap water and flush the salt solution down the sink.

WASTE
DISPOSE OF
PROPERLY

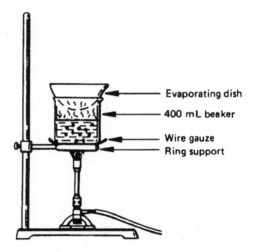

Figure 1.6 **Evaporation on a simple water bath**

- Evaporating dish
- 400 mL beaker
- Wire gauze
- Ring support

D. Filtration

The process of separating suspended insoluble solids from liquids by means of filters is called **filtration.** Insoluble solids, called **precipitates,** are formed during some chemical reactions. In the laboratory these precipitates are generally separated from the solutions by filtering them out on a paper filter. The liquid that passes through the filter paper is the **filtrate;** the solid precipitate remaining on the filter paper is the **residue.**

1. **Forming a Precipitate.** Fill a test tube about one-quarter full of lead(II) nitrate solution. Fill a second tube about one-half full of sodium iodide solution. These solutions contain lead(II) nitrate and sodium iodide, each dissolved in water. Pour the lead(II) nitrate solution into a 100 milliliter (mL) beaker. Slowly pour the sodium iodide solution into the beaker,

stir, and observe the results. The chemical reaction that occurred formed sodium nitrate and lead(II) iodide. One of these products is a yellow precipitate.

2. **Filtering the Products.** Prepare a filter as shown in Figure 1.7.

(a) Fold a circle of filter paper in half. Fold in half again and open out into a cone. Tear off one corner of the outside folded edge. The top edge of the cone which is to touch the glass funnel should not be torn.

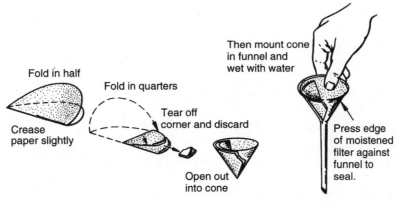

Figure 1.7 Folding and mounting filter paper

(b) Fit the opened cone into a short-stemmed funnel, placing the torn edge next to the glass. Wet with distilled water and press the top edge of the paper against the funnel, forming a seal. Use one of the setups suggested in Figure 1.8 for supporting the funnel. Then, stir the mixture of products in the small beaker with a stirring rod and slowly pour it down the stirring rod into the filter paper in the funnel (see Figure 1.9). Do not overfill the paper filter cone.

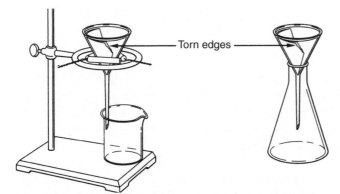

Figure 1.8 Support the filter with a ring stand or an Erlenmeyer flask

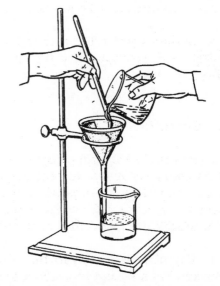

Figure 1.9 Pouring a solution down a stirring rod

3. **Identification of the Precipitate.** After the filtration is completed, compare the residue with the samples of solid sodium nitrate and lead(II) iodide provided by the instructor to determine which of these is the residue on the filter paper.

4. Use forceps to remove the filter paper with the precipitate and transfer it into the waste jar provided. Pour the filtrate into the waste bottle provided. Rinse the reaction beaker and the test tubes with water and pour the rinse solutions into the waste bottle.

5. Review the safety procedures for chemical waste disposal in the preface to determine the specific reasons for putting the waste into special containers rather than the trash and the sink.

REPORT FOR EXPERIMENT 1

Laboratory Techniques

A, B.   Laboratory Burners and Glassworking

Glassware shown in Figure 1.3

Articles	Instructor's Check and Comments
Straight tubes (2)	
Right-angle bends (2)	
Delivery tube (1)	
Buret tips (2)	(optional)
Stirring rod (1)	

Instructor's OK or grade on glass work _____

QUESTIONS AND PROBLEMS

1. Why is it necessary to turn off the gas with the gas cock rather than with the valve on the burner?

2. Why is air mixed with gas in the barrel of the burner before the gas is burned?

3. How would you adjust a burner which

   (a) has a yellow and smoky flame?

   (b) is noisy with a tendency to blow itself out?

4. Why are glass tubes and rods always fire-polished after cutting?

5. Explain briefly how to insert glass tubing into a rubber stopper.

6. Name the lubricant used for inserting glass tubing in rubber stoppers.

_____

C. Evaporation

Give the name and formula of the residue remaining after evaporation:

Name_____ Formula _____

D. Filtration

1. What is the name, formula, and color of the precipitate recovered by filtration?

Name _____ Formula _____ Color _____

2. Explain why the filter paper with the precipitate is collected in a jar instead of thrown into the trash can? (Refer to the section on waste disposal in Laboratory Rules and Safety Practices.)

3. Give the names and formulas of two compounds that must be present in the filtrate.

Name _____ Formula _____

Name _____ Formula _____

EXPERIMENT 2

Measurements

MATERIALS AND EQUIPMENT

Solids: sodium chloride (NaCl) and ice. Balance, ruler, thermometer, solid object for density determination, No. 1 or 2 solid rubber stopper.

DISCUSSION

Chemistry is an experimental science, and measurements are fundamental to most of the experiments. It is important to learn how to make and use these measurements properly.

The SI System of Units

The International System of Units (*Systeme Internationale, SI*) or metric system is a decimal system of units for measurements used almost exclusively in science. It is built around a set of units including the meter, the gram, and the liter and uses factors of 10 to express larger or smaller multiples of these units. To express larger or smaller units, prefixes are added to the names of the units. Deci, centi, and milli are units that are 1/10, 1/100, and 1/1000, respectively, of these units. The most common of these prefixes with their corresponding values expressed as decimals and powers of 10 are shown in the table below.

Prefix	Decimal Equivalent	Power of 10	Examples
deci (d)	0.1	10^{-1}	$1 \text{ dg} = 0.1 \text{ g} = 10^{-1} \text{ g}$
centi (c)	0.01	10^{-2}	$1 \text{ cm} = 0.01 \text{ m} = 10^{-2} \text{ m}$
milli (m)	0.001	10^{-3}	$1 \text{ mg} = 0.001 \text{ g} = 10^{-3} \text{ g}$
kilo (k)	1000	10^{3}	$1 \text{ km} = 1000 \text{ m} = 10^{3} \text{ m}$

Dimensional Analysis

It will often be necessary to convert from the American System of units to the SI system or to convert units within the SI system. Conversion factors are available from tables (see Appendix 4) or can be developed from the metric prefixes and their corresponding values as shown in the table above. Dimensional analysis, a problem-solving method with many applications in chemistry, is very valuable for converting one unit to another by the use of conversion factors. A review of using dimensional analysis for converting units is provided here. Study Aid 5 provides more help with this problem-solving tool.

Conversion Factors come from equivalent relationships, usually stated as equations. From each equivalence statement two conversion factors can be written in fractional form with a value of 1. For example:

Equivalence Equations	Conversion Factor #1	Conversion Factor #2
1 dollar = 4 quarters	$\dfrac{1 \text{ dollar}}{4 \text{ quarters}}$	$\dfrac{4 \text{ quarters}}{1 \text{ dollar}}$
1 lb = 453.6 g	$\dfrac{1 \text{ lb}}{453.6 \text{ g}}$	$\dfrac{453.6 \text{ g}}{1 \text{ lb}}$
1 mm = 10^{-3} m	$\dfrac{1 \text{ mm}}{10^{-3}\text{m}}$	$\dfrac{10^{-3}\text{m}}{1 \text{ mm}}$
1 ns = 10^{-9} s	$\dfrac{1 \text{ ns}}{10^{-9}\text{s}}$	$\dfrac{10^{-9}\text{s}}{1 \text{ ns}}$

The dimensional analysis method of converting units involves organizing one or more conversion factors into a logical series which cancels or eliminates all units except the unit(s) wanted in the answer.

For example: To convert 2.53 lb into milligrams (mg), the setup is:

$$(2.53 \text{ 1b})\left(\frac{453.6 \text{ g}}{1 \text{ lb}}\right)\left(\frac{1 \text{ mg}}{10^{-3}\text{g}}\right) = 1.15 \times 10^6 \text{ mg}$$

Note, that in completing this calculation, units are treated as numbers, **lb** in the denominator is canceled into **lb** in the numerator and **g** in the denominator is cancelled into **g** in the numerator. More examples of unit conversions can be found in Study Aid 5.

Although the SI unit of temperature is the Kelvin (K), the Celsius (or centigrade) temperature scale is commonly used in scientific work and the Fahrenheit scale is commonly used in this country. On the Celsius scale the freezing point of water is designated 0°C, the boiling point 100°C.

Precision and Accuracy of Measurements

Scientific measurements must be as **precise** as possible. This means that every measurement will include one uncertain or estimated digit. When making measurements we normally estimate between the smallest scale divisions on the instrument being used. Then, only the uncertain digit should vary if the measurement is repeated using the same instrument, even if it is repeated by someone else. The **accuracy** of a measurement or calculated quantity refers to its agreement with some known value. For example, we need to make two measurements, volume and mass, to determine the density of a metal. This experimental density can then be compared with the density of the metal listed in a reference such as the *Handbook of Chemistry and Physics*. High accuracy means there is good agreement between the experimental value and the known value listed in the reference. Not all measurements can be compared with a known value.

Random and Systematic Errors

The difference between the experimentally measured value of something and the accepted value of something is known as **the error.** For many of the experiments in this course, after you determine the error in your result, you may be required to find the percent error:

$$\text{Percent error} = \frac{\text{theoretical accepted value} - \text{experimentally determined value}}{\text{theoretical accepted value}} \times 100$$

There are two different types of error. **A random error** means that the error has an equal probablilty of being higher or lower than the accepted value. For example, a student measures the density of a quartz sample four times: (Accepted density value for quartz is 2.65 g/mL)

2.72 g/mL
2.55 g/mL Since two of the measured density values are below the mean
2.68 g/mL and two are above the mean, there is an **equal probability** of the
2.60 g/mL measurements being above or below the mean. This is a **random** error.
Since the mean density value is very close to the accepted value, the
Mean = 2.64 g/mL accuracy of the mean measurement is good. (the percent error is 0.38%)

The other type of error is a **systematic error.** This type of error occurs in the same direction each time (either always higher or always lower than the accepted value). For example, a student measures the boiling point of water four times (accepted temperature for the boiling point of water is 100.0° C.)

101.2° C
100.9° C Since all four of the measured temperature values are above the accepted
102.0° C value, the **error** is systematic. The mean value is 1.3% higher than the
101.0° C accepted value so the accuracy of these measurements is not as good as the
Mean = 101.3° C accuracy of the density of the measurements in the first example.

Precision and Significant Figures

When a measured value is determined to the highest precision of the measuring instrument, the digits in the measurement are called **significant digits** or **significant figures.**

Suppose we are measuring two pieces of wire, using the metric scale on a ruler that is calibrated in tenths of centimeters as shown in Figures 2.1a and b. One end of the first wire is placed at exactly 0.0 cm and the other end falls somewhere between 6.3 cm and 6.4 cm. Since the distance between 6.3 and 6.4 is very small, it is difficult to determine the next digit exactly. One person might estimate the length of the wire as 6.34 cm and another as 6.33 cm. The estimated digit is never ignored because it tells us that the ruler can be read to the 0.01 place. This measurement therefore has three significant figures (two certain and one uncertain figure).

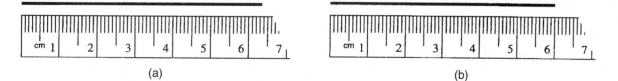

(a)   (b)

Figure 2.1

The second wire has a length which measures exactly 6 cm on the ruler as shown in Figure 2.1b. Reporting this length as 6 cm would be a mistake for it would imply that the 6 is an uncertain digit and others might record 5 or 7 as the measurement. Recording the measurement as 6.0 would also be incorrect because it implies that the 0 is uncertain and that someone else might estimate the length as 6.1 or 5.9. What we really mean is that, as closely as we can read it, the length is exactly 6 cm. So, we must write the number in such a way that it tells how precisely we can read it. In this example we can estimate to 0.01 cm so the length should be reported as 6.00 cm.

Significant Figures in Calculations

The result of multiplication, division, or other mathematical manipulation cannot be more precise than the least precise measurement used in the calculation. For instance, suppose we have an object that weighs 3.62 lb and we want to calculate the mass in grams. $(3.62 \text{ lb})\left(\dfrac{453.6 \text{ g}}{1 \text{ lb}}\right) = 1{,}642.032$ when done by a calculator. To report 1,642.032 g as the mass is absurd, for it implies a precision far beyond that of the original measurement. Although the conversion factor has four significant figures, the mass in pounds has only three significant figures. Therefore the answer should have only three significant figures; that is, 1,640 g. In this case the zero cannot be considered significant. This value can be more properly expressed as 1.64×10^3 g. For a more comprehensive discussion of significant figures see Study Aid 1.

Precise Quantities versus Approximate Quantities

In conducting an experiment it is often unnecessary to measure an exact quantity of material. For instance, the directions might state, "Weigh about 2 g of sodium sulfite." This instruction indicates that the measured quantity of salt should be 2 g plus or minus a small quantity. In this example 1.8 to 2.2 g will satisfy these requirements. To weigh exactly 2.00 g or 2.000 g wastes time since the directions call for approximately 2 g.

Sometimes it is necessary to measure an amount of material precisely within a stated quantity range. Suppose the directions read, "weigh about 2 g of sodium sulfite to the nearest 0.001 g." This instruction does not imply that the amount is 2.000 g but that it should be between 1.8 and 2.2 g and measured and recorded to three decimal places. Therefore, four different students might weigh their samples and obtain 2.141 g, 2.034 g, 1.812 g, and 1.937 g, respectively, and each would have satisfactorily followed the directions.

Temperature

The simple act of measuring a temperature with a thermometer can easily involve errors. Not only does the calibration of the scale on the thermometer limit the precision of the measurement, but the improper placement of the thermometer bulb in the material being measured introduces a common source of human error. When measuring the temperature of a liquid, one can minimize this type of error by observing the following procedures:

1. Hold the thermometer away from the walls of the container.

2. Allow sufficient time for the thermometer to reach equilibrium with the liquid.

3. Be sure the liquid is adequately mixed.

When converting from degrees Celsius to Fahrenheit or vice versa, we make use of the following formulas:

$$°C = \frac{(°F - 32)}{1.8} \text{ or } °F = (1.8 \times °C) + 32$$

Example Problem: Convert 70.0°F to degrees Celsius:

$$°C = \left(\frac{70.0°F - 32}{1.8}\right) = \frac{38.0}{1.8} = 21.11°C \text{ rounded to } 21.1°C$$

This example shows not only how the formula is used but also a typical setup of the way chemistry problems should be written. It shows how the numbers are used, but does not show the multiplication and division, which should be worked out by calculator. The answer was changed from 21.11°C to 21.1°C because the initial temperature, 70.0°F, has only three significant figures. The 1.8 and 32 in the formulas are exact numbers and have no effect on the number of significant figures.

Mass (Weight)

The directions in this manual are written for a 0.001 gram precision balance, but all the experiments can be performed satisfactorily using a 0.01 gram or 0.0001 gram precision balance. Your instructor will give specific directions on how to use the balance, but the following precautions should be observed:

1. The balance should always be "zeroed" before anything is placed on the balance pan. On an electronic digital balance, this is done with the "tare" or "T" button. Balances without this feature should be adjusted by the instructor.

2. Never place chemicals directly on the balance pan; first place them on a weighing paper, weighing "boat", or in a container. Clean up any materials you spill on or around the balance.

3. Before moving objects on and off the pan, be sure the balance is in the "arrest" position. When you leave the balance, return the balance to the "arrest" or standby position.

4. Never try to make adjustments on a balance. If it seems out of order, tell your instructor.

Volume

Beakers and flasks are marked to indicate only approximate volumes. Volume measurements are therefore made in a graduated cylinder by reading the point on the graduated scale that coincides with the bottom of the curved surface called the **meniscus** of the liquid (Figure 2.2). Volumes measured in this illustrated graduated cylinder are calibrated in 1 mL increments and should be estimated and recorded to the nearest 0.1 mL.

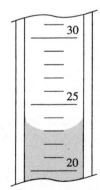

Figure 2.2 Read the bottom of the meniscus. The volume is 23.0 mL

Density

Density is a physical property of a substance and is useful in identifying the substance. **Density** is the ratio of the mass of a substance to the volume occupied by that mass; it is the mass per unit volume and is given by the equations

$$\text{Density} = d = \frac{\text{Mass}}{\text{Volume}} = \frac{m}{V} = \frac{g}{mL} \text{ or } \frac{g}{cm^3}$$

In calculating density it is important to make correct use of units and mathematical setups.

Example Problem: An object weighs 283.5 g and occupies a volume of 14.6 mL. What is its density?

$$d = \frac{m}{V} = \frac{283.5 \text{ g}}{14.6 \text{ mL}} = 19.4 \text{ g/mL}$$

Note that all the operations involved in the calculation are properly indicated and that all units are shown. If we divide grams by milliliters, we get an answer in grams per milliliter.

The volume of an irregularly shaped object is usually measured by the displacement of a liquid. An object completely submerged in a liquid displaces a volume of the liquid equal to the volume of the object.

Measurement data and calculations must always be accompanied by appropriate units.

PROCEDURE

Wear protective glasses.

Record your data on the report form as you complete each measurement, never on a scrap of paper which can be lost or misplaced.

A.   Temperature

Record all temperatures to the **nearest 0.1°C.**

1. Fill a 400 mL beaker half full of tap water. Place your thermometer in the beaker. Give it a minute to reach thermal equilibrium. Keeping the thermometer in the water and holding the tip of the thermometer away from the glass, read and record the temperature.

2. Fill a 150 mL beaker half full of tap water. Set up a ring stand with the ring and wire gauze at a height so the hottest part of the burner flame will reach the bottom of the beaker. Heat the water to boiling. Read and record the temperature of the boiling water, being sure to hold the thermometer away from the bottom of the beaker.

3. Fill a 250 mL beaker one-fourth full of tap water and add a 100 mL beaker of crushed ice. Without stirring, place the thermometer in the beaker, resting it on the bottom. Wait at least 1 minute, then read and record the temperature. Now stir the mixture for about 1 minute. If almost all the ice melts, add more. Holding the thermometer off the bottom, read and record the temperature. Save the ice-water mixture for Part 4.

4. Weigh approximately 5 g of sodium chloride and add it to the ice-water mixture. Stir for 1 minute, adding more ice if needed. Read and record the temperature. Dispose of the salt water/ice mixture in the sink.

B. Mass

Using the balance provided, do the following, recording all the masses to include one uncertain digit and all certain digits.

1. Weigh a 250 mL beaker.

2. Weigh a 125 mL Erlenmeyer flask.

3. Weigh a piece of weighing paper or a plastic weighing "boat."

4. Add approximately 2 g of sodium chloride to the weighing paper from step 3 and record the total mass. Calculate the mass of sodium chloride.

C. Length

Using a ruler, make the following measurements in centimeters; measure to the nearest uncertain digit.

1. Measure the length of the arrow on the right ⟶

2. Measure the external height of a 250 mL beaker.

3. Measure the length of a test tube.

D. Volume

Using the graduated cylinder most appropriate, measure the following volumes to the maximum precision possible, usually 0.1 mL. Remember to read the volume at the meniscus.

1. Fill a test tube to the brim with water and measure the volume of the water.

2. Fill a 125 mL Erlenmeyer flask to the brim with water and measure the volume of the water.

3. Measure 5.0 mL of water in a graduated cylinder and pour it into a test tube. With a ruler, measure the height (in cm) and mark the height with a marker.

4. Measure 10.0 mL of water in the graduated cylinder and pour it into a test tube like the one used in the previous step. Again, mark the height with a marker.

In the future, you will often find it convenient to estimate volumes of 5 and 10 mL simply by observing the height of the liquid in the test tube.

E. Density

Estimate and record all volumes to the highest precision, usually 0.1 mL. Make all weighing to the highest precision of the balance. Note that you must supply the units for the measurements and calculations in this section.

1. **Density of Water.**   Weigh a clean, dry 50 mL graduated cylinder and record its mass. (Graduated cylinders should never be dried over a flame.) Fill the graduated cylinder with distilled water to 50.0 mL. Use a medicine dropper to adjust the meniscus to the 50.0 mL mark. Record the volume. Reweigh and calculate the density of water.

2. **Density of a Rubber Stopper.**   Select a solid rubber stopper which is small enough to fit inside the 50 mL graduated cylinder. Weigh the dry stopper. Fill the 50 mL cylinder with tap water to approximately 25 mL. Read and record the exact volume. Carefully place the rubber stopper into the graduated cylinder so that it is submerged. Read and record the new volume. Calculate the volume and density of the rubber stopper.

3. **Density of a Solid Object.**   Obtain a solid object from your instructor. Record the sample code on the report form. Determine the density of your solid by following the procedure given in Part 2 for the rubber stopper. To avoid the possibility of breakage, incline the graduated cylinder at an angle and slide, rather than drop, the solid into it.

Return the solid object to your instructor.

NAME ____________________

SECTION __________ DATE __________

REPORT FOR EXPERIMENT 2

INSTRUCTOR ____________________

Measurements

A. Temperature

1. Water at room temperature ________ °C
2. Boiling point ________ °C
3. Ice water
   Before stirring ________ °C
   After stirring for 1 minute ________ °C
4. Ice water with salt added ________ °C

B. Mass

1. 250 mL beaker ________ g
2. 125 mL Erlenmeyer flask ________ g
3. Weighing paper or weighing boat ________ g
4. Mass of weighing paper/boat + sodium chloride ________ g
   Mass of sodium chloride (show calculation setup) ________ g

C. Length

1. Length of ⟶ ________ cm
2. Height of 250 mL beaker ________ cm
3. Length of test tube ________ cm

D. Volume

1. Test tube ________ mL
2. 125 mL Erlenmeyer flask ________ mL
3. Height of 5.0 mL of water in test tube ________ cm
4. Height of 10.0 mL of water in test tube ________ cm

E.   **Density**

1. **Density of Water**

Mass of empty graduated cylinder   _____

Volume of water   _____

Mass of graduated cylinder and water   _____

Mass of water (show calculation setup)   _____

Density of water (show calculation setup)   _____

2. **Density of a Rubber Stopper**

Mass of rubber stopper   _____

Initial volume of water in cylinder   _____

Final volume of water in cylinder (including stopper)   _____

Volume of rubber stopper (show calculation setup)   _____

Density of rubber stopper (show calculation setup)   _____

3. **Density of a Solid Object**

Number of solid object   _____

Mass of solid object   _____

Initial volume of water in graduated cylinder   _____

Final volume in graduated cylinder   _____

Volume of solid object (show calculation setup)   _____

Density of solid object (show calculation setup)   _____

QUESTIONS AND PROBLEMS

1. The directions state "weigh about 5 grams of sodium chloride". Give minimum and maximum amounts of sodium chloride that would satisfy these instructions.

2. Two students each measured the density of a quartz sample three times:

	Student A	Student B	
1.	3.20 g/mL	2.82 g/mL	The density found in the *Handbook*
2.	2.58 g/mL	2.48 g/mL	*of Chemistry and Physics* for quartz
3.	2.10 g/mL	2.59 g/mL	is 2.65 g/mL
mean	2.63 g/mL	2.63 g/mL	

(a) Which student measured density with the greatest precision? Explain your answer.

(b) Which student measured density with the greatest accuracy? Explain your answer.

(c) Are the errors for these students random or systematic? Explain.

Show calculation setups and answers for the following problems.

3. Convert 21°C to degrees Fahrenheit.   _____

4. Convert 101°F to degrees Celsius.   _____

5. An object is 9.6 cm long. What is the length in inches?   _____

6. An empty graduated cylinder weighs 82.450 g. When filled to 50.0 mL with an unknown liquid it weighs 110.810 g. What is the density of the unknown liquid?

_____

7. It is valuable to know that 1 milliliter (mL) equals 1 cubic centimeter (cm^3 or cc). How many cubic centimeters are in an 8.00 oz bottle of cough medicine? (1.00 oz = 29.6 mL)

_____

8. A metal sample weighs 56.8 g. How many ounces does this sample weigh? (1 lb = 16 oz)

_____

9. Convert 15 nm into km.

_____

EXPERIMENT 3

Preparation and Properties of Oxygen

MATERIALS AND EQUIPMENT

Solids: candles, magnesium (Mg) strips, manganese dioxide (MnO_2), fine steel wool (Fe), roll sulfur (S), wood splints. **Solution:** 9 percent hydrogen peroxide (H_2O_2). Deflagration spoon, pneumatic trough, 20 to 25 cm length rubber tubing, 25×200 mm ignition tube, five wide-mouth (gas-collecting) bottles, five glass cover-plates, Büchner funnel, heavy-wall filtering flask with side-arm tubulation, rubber suction tubing, filter paper to fit the Büchner funnel. **Demonstration supplies:** cotton, sodium peroxide (Na_2O_2); steel wool, 25×200 mm test tube; Hoffman electrolysis apparatus.

DISCUSSION

Oxygen is the most abundant and widespread of all the elements in the earth's crust. It occurs both as free oxygen gas and combined in compounds with other elements. Free oxygen gas is diatomic and has the formula O_2. Oxygen is found combined with more elements than any other single element, and it will combine with all the elements except some of the noble gases. Water is 88.9 percent oxygen by mass and the atmosphere is about 21 percent oxygen by volume. Oxygen gas is colorless and odorless, and is only very slightly soluble in water, a property important to its collection in this experiment.

Oxygen may be obtained by decomposing a variety of oxygen-containing compounds. Some of these are mercury(II) oxide (HgO, mercuric oxide), lead(IV) oxide (PbO_2, lead dioxide), potassium chlorate ($KClO_3$), potassium nitrate (KNO_3), hydrogen peroxide (H_2O_2), and water (H_2O).

In this experiment oxygen is produced by decomposing hydrogen peroxide, and five bottles of oxygen will be collected by the downward displacement of water. After collection, some of the physical and chemical properties of oxygen will be observed.

A. Decomposition of Hydrogen Peroxide to Generate Oxygen

Hydrogen peroxide decomposes very slowly at room temperature. The rate of decomposition is greatly increased by adding a catalyst, manganese dioxide. Although manganese dioxide contains oxygen, it is not decomposed under conditions of this experiment. These equations represent the changes that occur.

Word Equation: Hydrogen peroxide $\longrightarrow$ Water + Oxygen

Formula Equation: $2\,H_2O_2(aq) \xrightarrow{MnO_2} 2\,H_2O(l) + O_2(g)$

B. Collection of Oxygen

The oxygen is collected by a method known as the downward displacement of water. The gas is conducted from a generator to a bottle of water inverted in a pneumatic trough

(Figure 3.1). The oxygen, which is only very slightly soluble in the water, rises in the bottle and pushes the water down and out. Because oxygen is heavier than air, a glass plate is used to cover the opening of the bottle while it is inverted to a right-side-up position and placed on the benchtop until tested.

C. Properties of Oxygen

Like all kinds of matter, oxygen has both physical and chemical properties and you will observe both in this experiment. One outstanding and important chemical property of oxygen is its ability to support combustion. During combustion oxygen is consumed but does not burn and this ability to support combustion is one test for oxygen. Other substances (a wooden splint or a candle, for example) burn in oxygen producing a visible flame and heat. Compounds containing oxygen and one other element are known as oxides. Thus when elements such as sulfur, hydrogen, carbon, and magnesium burn in air or oxygen, they form sulfur dioxide, hydrogen oxide (water), carbon dioxide, and magnesium oxide, respectively. These chemical reactions may be represented by equations; for example:

Word Equation: Sulfur + Oxygen $\longrightarrow$ Sulfur dioxide

Formula Equation: $S(s) + O_2(g) \longrightarrow SO_2(g)$

See Study Aid 2 for a discussion of writing formulas and chemical equations.

PROCEDURE

A. and B.   Generation and Collection of Oxygen from Hydrogen Peroxide

Wear protective glasses.
Wash hydrogen peroxide off skin with water immediately.

1. Assemble the apparatus shown in Figure 3.1. It consists of a 250 mL Erlenmeyer flask, two-hole stopper, thistle tube, glass right-angle bend (Figure 1.3B), glass delivery tube with 135 degree bend (Figure 1.3C), and a 20–25 cm length of rubber tubing. The thistle tube should be at least 24 cm (~10 in.) long and be inserted in the rubber stopper so that there is about 3 mm (1/8 in.) clearance between the end of the tube and the bottom of the flask with the stopper in place. Remember to use glycerol when inserting the glass tubing into the rubber stopper and to hold the glass tubing close to the point of insertion.

2. Fill a pneumatic trough with water until the water level is just above the removable shelf. Attach a piece of rubber tubing to the overflow spigot on the trough and put it in the sink so the water will not spill over the edges of the trough onto the counter. Completely fill five wide-mouth bottles with water. Transfer each bottle to the pneumatic trough by covering its mouth with a glass plate, inverting it, and lowering it into the water. Remove the glass plate below the water level. Place two bottles on the shelf in the trough (over the holes), leaving the other three standing for transfer to the shelf when needed.

3. Using a spatula, put a pea-sized quantity of manganese dioxide (MnO_2) in the generator flask. Replace the stopper, stabilize the flask on the ring stand with a clamp, and make sure that all glass-rubber connections are tight. Add 25 mL of water to the flask through the thistle tube. Make sure that the end of the thistle tube is covered with water (to prevent escape of oxygen gas through the thistle tube).

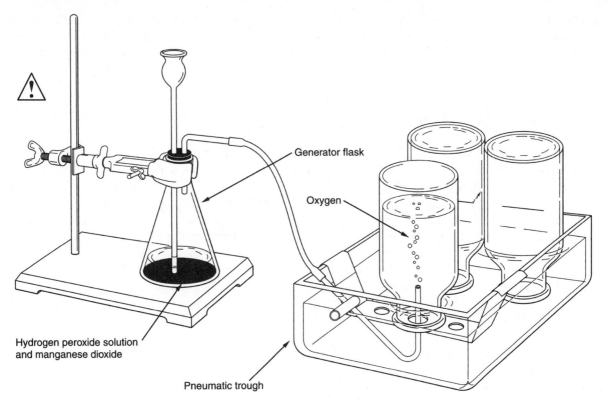

Figure 3.1   Preparing oxygen by decomposing hydrogen peroxide

4. Using a 50 mL graduated cylinder, measure about 50 mL of 9 percent hydrogen peroxide solution.

 Reminder: If hydrogen peroxide gets on your skin, wash it off promptly with water.

To start the generation of oxygen, pour 5 to 10 mL of the peroxide solution into the thistle tube. If all the peroxide solution does not run into the generator, momentarily lift the delivery tube from the water in the trough. Immediately replace the end of the delivery tube under water and into the mouth of the first bottle to collect the gas. When one bottle is filled with gas, immediately start filling the next bottle. Continue generating oxygen by adding an additional 5 to 10 mL portion of hydrogen peroxide whenever the rate of gas production slows down markedly.

5. Cover the mouth of each gas-filled bottle with a glass cover-plate before removing it from the water. Store each bottle mouth upward without removing the glass plate; the oxygen will not readily escape since it is slightly more dense than air. **Note which bottle of gas was collected first** and continue until a total of five bottles of gas have been collected.

6. Allow the reaction to go to completion while you continue with the testing of the oxygen you collected. If you have any unreacted H_2O_2 remaining in the graduated cylinder, return it to the special bottle marked "9% unreacted H_2O_2." When you have completed the rest of this experiment, pour the material in the generator into the vacuum flask through the Büchner funnel setup (see Figure 3.2) for waste MnO_2 disposal. Rinse the generator with water. Occasionally the filter paper will need to be changed and the filter flask emptied into the sink.

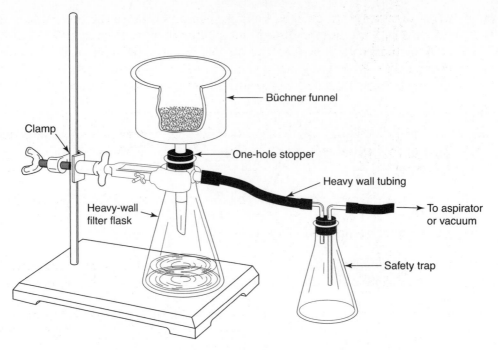

Figure 3.2   Büchner funnel-vacuum flask setup

C.   Properties of Oxygen

Each of the following tests (except C.6) is conducted with a bottle of oxygen and, for comparison, with a bottle of air. Record your observations on the report form.

   1. The **glowing splint test** is often used to verify the identity of oxygen. Ignite a wood splint, blow out the flame, and insert the still-glowing splint into the first bottle of oxygen collected. Repeat with a bottle of air. To ensure having a bottle of air, fill the bottle with water and then empty it, thus washing out other gases that may be present.

   2. Take a small lump of sulfur in a deflagrating spoon, a bottle of oxygen, and a bottle of air to the fume hood. Light the burner in the fume hood and direct the flame directly into the spoon containing the sulfur. First the sulfur gets dark and melts, then it begins to burn with a blue flame that is barely visible. Lower the burning sulfur alternately into a bottle of oxygen and a bottle of air and compare combustions. Quench the excess burning sulfur in a beaker of water.

   3. Stand a small candle (no longer than 5 cm) on a glass plate and light it. Lower a bottle of oxygen over the burning candle, placing the mouth of the bottle on the glass plate. **Measure and record the time,** in seconds, that the candle continues to burn. Repeat with a bottle of air. Note also the difference in the brilliance of the candle flame in oxygen and in air. Return the unused portion of the candle to the reagent shelf.

   4. Invert a bottle of oxygen, covered with glass plate, and place it mouth to mouth over a bottle of air. Then remove the glass plate from between the bottles and allow them to stand mouth to mouth for 3 minutes. Cover each bottle with a glass plate and set the bottles down, mouths upward. Test the contents of each bottle by inserting a glowing splint.

5. Pour 25 mL of water into the fifth bottle of oxygen and replace the cover. Place the bottle close to (within 5 or 6 cm) the burner. Take a loose, 4 or 5 cm wad of steel wool (iron) in the crucible tongs and momentarily heat it in the burner flame until some of the steel wool first begins to glow. Immediately lower the glowing metal into the bottle of oxygen. (It is essential that some of the steel wool be glowing when it goes into the oxygen.) Repeat, using a bottle of air.

> **NOTE:** The 25 mL of water is to prevent breakage if the glowing steel wool is accidentally dropped into the bottle.

6. A small strip of magnesium ribbon will be burned next. Read the following precautions before proceeding. *Do not put burning magnesium into a bottle of oxygen.* There is enough oxygen in air for this reaction to proceed vigorously.

 Do not look directly at the burning magnesium ribbon. It is very bright and the light includes considerable ultraviolet light, which can cause damage to the retina of the eye.

Take a 2 to 5 cm strip of magnesium metal in a pair of crucible tongs and ignite it by heating it in the burner flame. After the burning is over, put the product on the Ceramfab plate and compare it to the metal from which it was produced.

D. Instructor Demonstrations (Optional)

1. **Sodium Peroxide as a Source of Oxygen.** Spread some cotton on the bottom of an evaporating dish and sprinkle a small amount (less than 1 g) of fresh sodium peroxide on it. Sprinkle a few drops of water on the peroxide. Spontaneous combustion of the cotton will occur.

2. **Approximate Percentage of Oxygen in the Air.** Push a small wad of steel wool to the bottom of a 25×200 mm test tube. Wet the steel wool by covering with water; pour out the surplus water; and place the tube, mouth downward, in a 400 mL beaker half full of water. After the oxygen in the trapped air has reacted with the steel wool—at least three days are needed for complete reaction—adjust the water levels inside and outside the tube to the same height. Cover the mouth of the tube, remove from the beaker, and measure the volume of water in the tube. Alternatively, the height of the water column may be measured (in millimeters) without removing the tube from the beaker. The volume of water in the tube is approximately equal to the volume of oxygen originally present in the tube of air.

$$\% \text{ oxygen} = \left(\frac{\text{Volume of water in tube}}{\text{Volume of tube}} \right)(100)$$

or

$$\% \text{ oxygen} = \left(\frac{\text{Height of water column}}{\text{Length of tube}} \right)(100)$$

3. **Decomposition of Water.** Set up the Hoffman electrolysis apparatus, as shown in Figure 3.3. The solution used in the apparatus should contain about 2 mL of sulfuric acid per 100 mL of water. Direct current may be obtained from several 1.5 volt type A cells connected in series or from some other D.C. source.

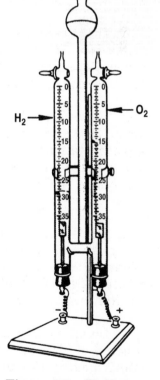

Figure 3.3 Hoffman electrolysis apparatus

REPORT FOR EXPERIMENT 3

Preparation and Properties of Oxygen

A. and B.   Generation and Collection of Oxygen

1. What evidence did you observe that oxygen is not very soluble in water?

2. What is the source of oxygen in the procedure you used?

    Name _____ Formula _____

3. What purpose does the manganese dioxide serve in this preparation of oxygen?

4. What gas was in the apparatus before you started generating oxygen? Where did it go?

5. What is different about the composition of the first bottle of gas collected compared to the other four?

6. Why are the bottles of oxygen stored with the mouth up?

7. (a) What is the symbol of the element oxygen?   _____

    (b) What is the formula for oxygen gas?   _____

8. Which of the following formulas represent oxides? (Circle) MgO, $KClO_3$, SO_2, MnO_2, O_2,   NaOH, PbO_2, Na_2O_2

9. Write the word and formula equations for the preparation of oxygen from hydrogen peroxide.

   Word Equation:

   Formula Equation:

10. What substances, other than oxygen, are in the generator when the decomposition of H_2O_2 is complete?

C. Properties of Oxygen

1. Write word equations for the chemical reactions that occurred. (See Study Aid 2.)

   C.1. Combustion of wood. Assume carbon is the combustible material.

   C.2. Combustion of sulfur.

   C.5. Combustion of steel wool (iron). (Call the product iron oxide.)

   C.6. Combustion of magnesium.

2. Write formula equations for these four chemical reactions.

   C.1. (CO_2 is the formula for the oxide of carbon that is formed.)

   C.2. (SO_2 is the formula for the oxide of sulfur that is formed.)

C.5. (Fe_3O_4 is the formula for the oxide of iron that is formed.)

C.6. (MgO is the formula for the oxide of magnesium that is formed.)

3. Combustion of a candle.

  (a) Number of seconds that the candle burned in the bottle of oxygen.   _____

  (b) Number of seconds that the candle burned in the bottle of air.   _____

  (c) Explain this difference in combustion time.

  (d) Is it scientifically sound to conclude that all the oxygen in the bottle was reacted when the candle stopped burning? Explain.

4. What were the results of the experiment in which a bottle of oxygen was placed over a bottle of air? Explain the results.

5. (a) Describe the material that is formed when magnesium is burned in air.

  (b) What elements are in this product?

6. (a) What is your conclusion about the rate or speed of a chemical reaction with respect to the concentration of the reactants—for example, a combustion in a high concentration of oxygen (pure oxygen) compared to a combustion in a low concentration of oxygen (air)?

   (b) What evidence did you observe in the burning of sulfur to confirm your conclusion in 6(a)?

EXPERIMENT 4

Preparation and Properties of Hydrogen

MATERIALS AND EQUIPMENT

Solids: strips of copper, magnesium, and zinc; sodium metal; steel wool; mossy zinc; wood splints. **Solutions:** dilute (6 M) acetic acid ($HC_2H_3O_2$), 0.1 M copper(II) sulfate ($CuSO_4$), dilute (6 M) hydrochloric acid (HCl), dilute (3 M) phosphoric acid (H_3PO_4), 9 M sulfuric acid (H_2SO_4) and dilute (3 M) sulfuric acid (H_2SO_4), phenolphthalein solution. Pneumatic trough, five wide-mouth (gas-collecting) bottles, pH paper.

DISCUSSION

Hydrogen, having atomic number 1 and atomic mass 1.008 amu, is the simplest element. It is the ninth most abundant element in the earth's crust (about 0.9 percent by mass). At ordinary temperatures and pressures it is a gas, composed of diatomic molecules, H_2, and is only very slightly soluble in water. Hydrogen is usually found combined with other elements. Water is the most common and probably the most important compound of hydrogen. Hydrogen will not support combustion, but in the presence of oxygen it burns readily to form water:

$$2\,H_2(g) + O_2(g) \longrightarrow 2\,H_2O(g)$$

This reaction is used as a simple test for hydrogen, for mixtures of hydrogen and air (or oxygen) burn explosively with a distinctive "popping" or "barking" sound.

A.   Methods of Preparing Hydrogen

There are several ways of producing hydrogen gas. Two methods are used in this experiment. Both involve the formation of hydrogen from aqueous solutions containing hydrogen ions (H^+). The concentration of hydrogen ions in solution is often expressed as pH. pH is a numerical scale corresponding to the acidity of a solution. Pure water at 25°C is neither acidic nor basic and is given a pH of 7 on the pH scale. pH values less than 7 indicate an acidic solution; the lower the number, the more acidic the solution. pH values greater than 7 indicate a basic solution; the larger the number the more basic the solution.

> pH < 7 = acid solutions
>
> pH = 7 = neutral solution
>
> pH > 7 = basic solutions

It is possible to determine whether a solution is acidic or basic by using acid–base indicators such as litmus or phenolphthalein. However, these indicators do not give very precise information. It is also possible to get more precise hydrogen ion concentrations using pH paper strips impregnated with several indicators, or with an instrument called a pH meter. In this experiment you will measure the pH of several acids and compare the relationship of their pH values to the strength of the acids and the rate of hydrogen formation.

1. **Active Metal with Water.** Several of the most active metals—such as lithium, sodium, potassium, rubidium, cesium, and calcium—will react with and displace hydrogen from cold water; magnesium will react with hot water. An example of such a reaction is

$$\text{Sodium} + \text{Water} \longrightarrow \text{Sodium hydroxide} + \text{Hydrogen}$$

$$2\,Na(s) + 2\,H_2O(l) \longrightarrow 2\,NaOH(aq) + H_2(g)$$

Besides hydrogen, this reaction also produces sodium hydroxide, which causes the solution to become basic (alkaline, pH > 7). Basic solutions can be differentiated from neutral or acidic solutions by using acid-base indicators such as litmus or phenolphthalein, pH paper, or a pH meter.

2. **Active Metal with Dilute Acid.** In general, metals that are more active than hydrogen will react with dilute acids to produced hydrogen gas and a salt. For example, the salts produced in the following reactions are magnesium chloride, zinc sulfate, and zinc phosphate, respectively.

$$Mg(s) + 2\,HCl(aq) \longrightarrow MgCl_2(aq) + H_2(g)$$

$$Zn(s) + H_2SO_4(aq) \longrightarrow ZnSO_4(aq) + H_2(g)$$

$$3\,Zn(s) + 2\,H_3PO_4(aq) \longrightarrow Zn_3(PO_4)_2(aq) + 3\,H_2(g)$$

The strong oxidizing acids, such as nitric acid (HNO_3) and concentrated sulfuric acid, also react with metals but do not produce hydrogen gas.

B. Collection of Hydrogen

Hydrogen, like other gases such as oxygen, which are insoluble in water, is collected by the method known as the downward displacement of water. The gas is conducted from a generator to a bottle of water inverted in a pneumatic trough (Figure 4.1). The insoluble hydrogen rises in the bottle and pushes the water down and out. Because hydrogen gas is lighter than air, bottles containing hydrogen are stored upside down until the gas is ready to be tested.

C. Properties of Hydrogen

During this experiment you will collect several bottles of hydrogen and observe some of its physical and chemical properties. The physical properties tested are the density of hydrogen relative to air and its ability to spontaneously mix with other gases (diffusion). When observing the chemical properties of hydrogen be sure to keep in mind that there is a distinction between combustion and the ability to support combustion.

We can observe three different situations: (1) A flame inserted into a bottle of pure hydrogen will go out, because there is no oxygen to support combustion. (2) A flame inserted into a mixture of air and hydrogen will set off a very rapid combustion (burning) of the hydrogen, causing a small explosion. (3) A flame brought to the mouth of a bottle of pure hydrogen will cause the hydrogen to burn, but only at the mouth, where the hydrogen is in contact with the air. Situation (3) is the hardest to detect because the reaction is not explosive, there is little noise, and hydrogen burns with an almost colorless flame.

PROCEDURE

 PRECAUTIONS:

1. **Wear protective glasses.**

2. Mixtures of hydrogen and air may explode when ignited. Keep the burner away from the generator tube and the delivery tube. Wrap the generator with a towel to prevent flying glass.

A. Preparing Hydrogen from Water

Record your observations on the report form.

 Do not put your head over the tube while the sodium is reacting.

1. Fill a standard test tube half full of water and place it in the test tube rack. Obtain a piece of sodium (no larger than a 4 mm cube) from your instructor and place it on a piece of filter paper; do not touch the sodium with your fingers. Fold the filter paper over the sodium and press out the kerosene, noting how soft the sodium metal is. (Sodium is stored in kerosene because it reacts rapidly with water or with the oxygen in air.) Pick up the sodium with tweezers or tongs and, holding it at arm's length, drop it into the test tube.

Immediately bring a burning splint to the mouth of the test tube and observe the results. When the reaction has ceased, use a clean stirring rod to place drops of the solution on pieces of red and blue litmus paper to determine whether the solution is acidic or basic. Acids turn litmus red, bases turn litmus blue. Add a drop of phenolphthalein to the solution in the test tube. A bright pink color indicates the presence of a base. Measure the pH of the solution by placing drops of the solution on pH paper and comparing the color to the standards provided.

2. Empty the solution in the test tube into the sink and flush generously with water.

B. Preparing Hydrogen from Acids

1. **Using Various Metals.**   Set up four test tubes in a rack. Place one small strip of zinc into the first tube. In like manner place samples of copper, steel wool (iron), and magnesium, in this order, into the other three tubes. In rapid succession add a few milliliters of dilute (6 M) hydrochloric acid to each test tube. Note whether gas is evolved in each case and also note its relative rate of evolution. Test any gas evolved for evidence of hydrogen by promptly bringing a burning splint to the mouth of the test tube.

Carefully pour the liquids in the test tubes into the sink keeping the metal strips in the test tubes. Dispose of the unreacted metal strips by rinsing with water and putting them into the container provided. Waste solids are never put into the sink.

2. **Using Various Acids.**   In the preceding section you found that zinc is one of the metals capable of releasing hydrogen from hydrochloric acid. In this section you will compare the relative ease with which zinc displaces hydrogen from a variety of acids. The stronger acids react at a faster rate than the weaker acids.

Set up four test tubes in a rack. Place several milliliters of dilute (6 M) hydrochloric acid into the first tube, dilute (6 M) acetic acid into the second, dilute (3 M) sulfuric acid into the third, and dilute (3 M) phosphoric acid into the fourth. Measure the pH of each acid using pH paper as before. Now drop a small strip of zinc into each of the four tubes. The variation in the rates of evolution of hydrogen gas is a measure of the relative strengths of the acids. Let the reactions proceed for three minutes before making your evaluation.

 Carefully pour the liquids into the sink, keeping the zinc in the test tube. Rinse the unreacted zinc with water and dispose of the zinc in the container provided, not in the sink or the trash.

C. Generation and Collection of Hydrogen

1. Assemble the generator shown in Figure 4.1, using a wide-mouth bottle or a 250 mL Erlenmeyer flask, a 2-hole rubber stopper equipped with a thistle tube reaching to within 1 cm of the bottom of the bottle, and a delivery tube. Clamp the generator to the ring stand. All connections must be airtight.

2. Fill the pneumatic trough with water until the water level is just above the removable shelf. Attach a piece of rubber tubing to the overflow spigot on the trough and put it in the sink so the water will not spill onto the lab bench.

3. Place approximately 10 g of mossy zinc in the bottle; add 2 mL of 0.1 M copper(II) sulfate solution (as a catalyst) and 50 mL of water and make sure that the bottom end of the thistle tube is under water. Fill four wide-mouth bottles with water and invert them in the pneumatic trough with two over the holes in the shelf. Put the delivery tube into the opening of one of the bottles.

4. When you are ready to collect the gas, pour about 4–5 mL of 9 M sulfuric acid, H_2SO_4, through the thistle tube. Add more acid in 4–5 mL increments, as needed, to keep the reaction going. As each bottle becomes filled with gas, place a glass plate over its mouth while the bottle is still under water; then remove the bottle from the water and store it mouth downward without removing the glass plate.

5. Set aside—but do not discard—the first bottle of gas collected, and fill the other three with hydrogen.

6. Since no more hydrogen gas is needed, fill the generator with water to dilute the acid and quench the reaction. Open the generator, pour out the diluted acid into the sink and flush generously with water. Be careful. The leftover zinc should NOT go into the sink. Rinse the remaining zinc thoroughly with water, and return this zinc to the container provided.

D. Properties of Hydrogen

NOTE: Unless otherwise directed, keep the bottles mouth downward while performing the following tests.

1. Raise the first bottle of gas collected a few inches straight up from the table, and immediately apply a burning splint to the mouth of the bottle.

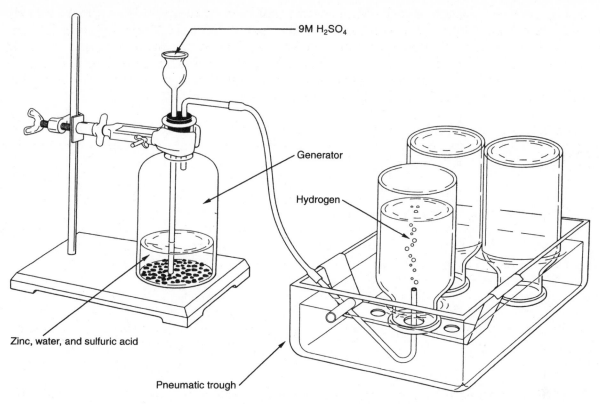

9M H₂SO₄

Generator

Hydrogen

Zinc, water, and sulfuric acid

Pneumatic trough

Figure 4.1   Preparing hydrogen from zinc and sulfuric acid

2. Raise the second bottle straight up a few inches from the table. Then without delay slowly insert a burning splint halfway into the bottle. Continue with the insertion of the splint even though you hear a muffled report as the flame approaches the mouth of the bottle. Slowly with-draw the splint until the charred end is in the neck of the bottle; hold the splint there a few seconds, but do not withdraw it completely. Repeat inserting and withdrawing the splint several times. Record your observations.

3. Place the third bottle mouth upward and remove the glass plate. After one minute, bring a burning splint to its mouth. Record your observations.

4. Keeping the cover plate over its mouth, place the fourth bottle of hydrogen (still upside down) mouth to mouth over a bottle of air. Then remove the cover plate from between them. Let the two bottles remain in this position for three minutes, then replace the cover plate between them. Leaving the cover plate on the mouth of the lower bottle, raise the top bottle straight up, at least 6 inches, and immediately bring a burning splint to its mouth. Turn the lower bottle mouth downward, with the cover plate in place. Bring a burning splint to the mouth of this bottle as you lift it straight up. Compare the results.

NAME ______________________________

SECTION __________ DATE ___________

REPORT FOR EXPERIMENT 4

INSTRUCTOR _____________________

Preparation and Properties of Hydrogen

A. Preparing Hydrogen from Water

1. Describe what you observed when sodium was dropped into water.

2. Describe what you observed when a flame was brought to the mouth of the test tube.

3. (a) What color did the litmus papers turn? ___________

   (b) What color did the phenolphthalein turn? ___________

   (c) What was the pH of the solution? ___________

4. Did the reaction make the solution acidic or basic? ___________

5. Complete and balance the following word and formula equations:

   Sodium + Water $\longrightarrow$

   $Na + H_2O \longrightarrow$

B. Preparing Hydrogen from Acids

1. (a) Write the symbols of the metals that reacted with dilute hydrochloric acid.

   ____

   (b) Write the symbols of the four metals in order of decreasing ability to liberate hydrogen from hydrochloric acid.

   ___________ ___________ ___________ ___________

2. Write the formulas of the four acids in order of decreasing strength based on the rates at which they displace hydrogen when reacting with zinc. Record the pH below each formula.

formula   _____   _____   _____   _____

pH   _____   _____   _____   _____

C.   Generation and Collection of Hydrogen

1. Why was water added to cover the bottom of the thistle tube?

2. What do we call this method of collecting a gas?

3. What physical property of hydrogen, other than that it is less dense than water, allows it to be collected in this manner?

D.   Properties of Hydrogen

1. What happened when the splint was brought to the mouth of the first bottle of gas collected?

2. Describe fully the results of testing the second bottle of gas.

3. (a) Is hydrogen combustible?   _____

   (b) What evidence of this do you have from testing the second bottle of gas?

4. (a) Does hydrogen support combustion?   _____

   (b) What evidence do you have of this from testing the second bottle of gas?

5. What compound was formed during the testing of the first and second bottles of gas?

_____

6. Why did the first bottle of gas behave differently from the second bottle?

7. (a) What happened when the splint was brought to the mouth of the third bottle of gas?

(b) How do you account for this?

8. When testing the fourth bottle:

(a) What was the result with the top bottle?

(b) What was the result with the lower bottle?

(c) How do you account for these results?

9. Complete the following word equations:

Zinc + Sulfuric acid $\longrightarrow$

Magnesium + Hydrochloric acid $\longrightarrow$

Hydrogen + Oxygen $\longrightarrow$

10. Complete and balance the following corresponding formula equations:

$Zn(s) + H_2SO_4(aq) \longrightarrow$

$Mg(s) + HCl(aq) \longrightarrow$

$H_2(g) + O_2(g) \longrightarrow$

QUESTIONS AND PROBLEMS

1. What does pH measure?

2. Which gave better evidence of acid strength, evolution of hydrogen or pH? Explain your answer.

EXPERIMENT 5

Calorimetry and Specific Heat

MATERIALS AND EQUIPMENT

Styrofoam cups, 6 oz; thermometers, metal samples, test tube with a diameter of at least 22 mm, bunsen burner, wire gauze, 400 ml beaker, cardboard cut into 4" squares with a small hole in the middle for a thermometer.

DISCUSSION

Calorimetry is the science of measuring a quantity of heat. Heat is a form of energy associated with the motion of atoms or molecules of a substance. Heat (often represented as "q") is measured in energy units such as joules or calories. Temperature (often represented as "t") is measured in degrees (usually Celsius). The measurement of temperature is already familiar to you. The same temperature is obtained for the water in a lake and for a thermos of water taken from the lake. But the heat content of the whole lake is much more than the heat content in that thermos of water even though both are exactly the same temperature.

Temperature and heat are related to each other by the specific heat ($sp\ ht$) of a substance, defined as the quantity of heat needed to raise one gram of a substance by one degree Celsius (J/g°C). The relationship between quantity of heat (q), specific heat ($sp\ ht$), mass (m) and temperature change (Δt) is mathematically expressed by the equation:

$$q = (m)(sp\ ht)(\Delta t) \ \text{ or } \ \text{Joules} \ = (g)\left(\frac{J}{g°C}\right)(\Delta°C) = (\cancel{g})\left(\frac{J}{\cancel{g}°\cancel{C}}\right)(\Delta°\cancel{C}) = J$$

Since the mass and temperature can be measured by a balance and a thermometer, respectively, q can be calculated if the $sp\ ht$ for a substance is known. Also, sp ht can be calculated if the heat content (q) of the substance is known. The amount of heat needed to raise the temperature of 1 g of water by 1 degree Celsius is the basis of the calorie. Thus, the specific heat of water is exactly 1.00 cal/g°C. The SI unit of energy is the joule and it is related to the calorie by 1 calorie = 4.184 J. Thus, the specific heat of water is also 4.184 J/g°C. The specific heat of a substance relates to its capacity to absorb heat energy. The higher the specific heat of a substance the more energy required to change its temperature.

The specific heat of metals generally varies with their atomic masses. You will see this relationship later when the data in Table 5.1 is graphed. For this data, atomic mass is the independent variable and specific heat is the dependent variable because we are examining how specific heat changes as a function of atomic mass. For a review of variables and graphing techniques, see Study Aid 3.

In this experiment, we will use calorimetry to determine the specific heat of a metal. Heat energy is transferred from a hot metal to water until the metal and the water have reached the same temperature. This transfer is done in an insulated container to minimize heat losses to

Table 5.1 Specific Heat of Selected Metals

Name of Metal	Atomic Mass, amu	Specific Heat, J/g°C
Aluminum	26.98	0.900
Copper	63.55	0.385
Gold	197.0	0.131
Iron	55.85	0.451
Lead	207.2	0.128
Silver	107.9	0.237
Tin	118.7	0.222

the surroundings. We then make the assumption that all the heat lost by the metal (q_x) was absorbed by the water and is equal to the heat gained by the water, (q_w). Since we know the specific heat of water, we have all the variables needed to calculate q_w using the equation:

$$q_w = (m_w)(sp\ ht_w)(\Delta t_w)$$

Since q_w is equal to q_x we can say that

$$q_w = q_x = (m_x)(sp\ ht_x)(\Delta t_x)$$

This relationship can be used to calculate $sp\ ht_x$ of a metal because both m_x and Δt_x can be measured.

Sample Calculation:

A metal sample weighing 68.3820 g was heated to 99.0°C, then quickly transferred into a styro-foam calorimeter containing 62.5515 g of distilled water at a temperature of 18.0°C. The temperature of the water in the styrofoam cup increased and stabilized at 20.6°C. Calculate the $sp\ ht_x$ and identify the metal using 4.184 J/g°C for the $sp\ ht_w$.

$$\Delta t_w = 20.6°C - 18.0°C = 2.6°C$$
$$q_w = (m_w)(sp\ ht_w)(\Delta t_w)$$
$$= (62.5515\,g)(4.184\,J/g°C)(2.6°C)$$
$$= 680\,J \quad \text{(heat absorbed by the water)}$$

Let x be the metal. Then, since all of the heat absorbed by the water came from the hot metal, we can say that

$$q_w = q_x = (m_x)(sp\ ht_x)\Delta t_x$$
$$\Delta t_x = 99.0°C - 20.6°C = 78.4°C$$
$$680\,J = (68.3820\,g)(sp\ ht_x)(78.4°C)$$
$$sp\ ht_x = 0.13\,J/g°C$$

Refer to Table 5.1 and determine that the unknown metal is lead or gold.

PROCEDURE: TRIAL 1

Wear protective glasses.

 No waste generated by this experiment.

1. Weigh a dry metal sample and record the mass and the name of the metal on the report form.

2. Carefully slide the sample into a large dry test tube and put a thermometer beside it in the test tube.

3. Attach the test tube to a ring stand and place it into an empty 400 mL beaker as shown in Figure 5.1. Be sure the height of the beaker is adjusted so the hottest part of the burner flame will be on the bottom of the beaker. Do not heat the dry beaker while making this adjustment. The bottom of the test tube should be at least one-half inch above the bottom of the beaker.

4. Fill the beaker with tap water so the height of the water in the beaker is about two inches higher than the top of the metal sample. There should be no water inside the test tube.

5. Begin heating the water in the beaker and continue with the next step(s). As you are working, check the water and note when it starts to boil. Turn down the burner but keep the water gently boiling. Do not do step 11 until the water has been boiling for about ten minutes and the temperature in the test tube has stabilized.

6. Nest two dry styrofoam cups together, weigh them, and record the mass on your report form.

7. Take another metal sample similar to the one you are heating. It does not matter if it is not the same metal. It is a "stand in" for the metal sample you are heating and will not be used during the experiment. Put this "stand in" into the styrofoam cup and add enough distilled water to cover the metal by no more than one-half inch.

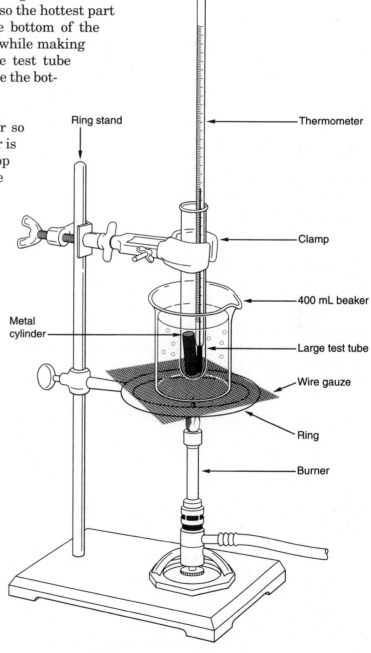

Figure 5.1   The boiler

– 45 –

8. Remove the "stand in" and weigh and record the mass of the styrofoam cups plus the water that you added.

9. Take a cardboard cover for the styrofoam cup and insert a thermometer through the hole. The nested cups with the cardboard cover and thermometer are referred to as a calorimeter, Figure 5.2. If you just leave the calorimeter on the benchtop it might fall over and break the thermometer so put the whole setup into a small beaker to stabilize it. The cardboard cover must rest directly on top of the styrofoam cup and not on the beaker.

10. Measure and record the temperature of the water in the styrofoam cup. Leave the thermometer in the cover until you are ready to transfer the hot metal into the calorimeter.

11. After the water in the beaker has been boiling for 10 minutes and the temperature inside the test tube with the metal has been stable for 5 minutes record the temperature on your report form. Remove the thermometer from the test tube and set it aside so it does not get mixed up with the thermometer used in the calorimeter.

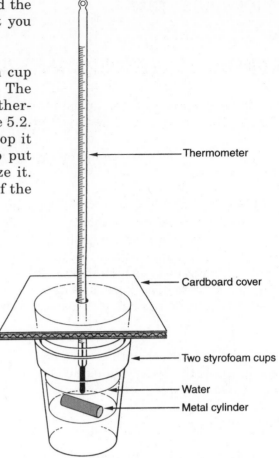

Figure 5.2   The calorimeter (see #9)

12. Now, you are going to transfer the metal from the test tube to the water in the calorimeter. It is important that the transfer take place quickly and carefully to minimize heat loss to the surroundings and to avoid splashing. Remove the cardboard cover and thermometer from the calorimeter. Loosen the clamp on the boiler ring stand, lift the clamp and test tube out of the boiler, and quickly slide the metal into the water in the calorimeter.

13. Immediately, put the cardboard cover with the thermometer back on the styrofoam cup. Stir gently for 2–3 minutes while monitoring the temperature. Record the temperature after it has remained constant for about one minute.

14. Unless instructed otherwise, repeat the experiment with Trial 2.

PROCEDURE: TRIAL 2

15. Repeat this experiment using the same metal sample, but this time use colder distilled water (5–10°C). Do not record the initial temperature of the cold water in the calorimeter until immediately before you add the hot metal. Fill out the report form in the column labeled Trial 2.

16. Look up and record the theoretical value for the specific heat for your metal sample. This can be found in Table 5.1 or in the *Handbook of Chemistry and Physics* if it is a metal not listed in the table.

REPORT FOR EXPERIMENT 5

Calorimetry and Specific Heat

Measurements and Calculations

	TRIAL 1	TRIAL 2
1. Mass of metal sample	_____	_____
2. Mass of calorimeter (styrofoam cups)	_____	_____
3. Mass of styrofoam cups + water	_____	_____
4. Mass of water (show calculation setup)	_____	_____
5. Initial water temperature	_____	_____
6. Temperature of heated metal sample	_____	_____
7. Final temperature of water and metal	_____	_____
8. Change in temperature, Δt_w, of the water in the calorimeter (show calculation setup)	_____	_____
9. Change in temperature, Δt_x, of the metal sample (show calculation setup)	_____	_____
10. Specific heat ($sp\ ht_w$) of water	$4.184\ J/g°C$	$4.184\ J/g°C$
11. Heat (q_w) gained by water (show calculation setup)	_____	_____
12. Heat (q_x) lost by metal sample	_____	_____
13. Specific heat ($sp\ ht_x$) of the metal (experimental value) (show calculation setup)	_____	_____

14. Name of metal _____   Theoretical Specific Heat _____

QUESTIONS AND PROBLEMS

1. Why is it important for there to be enough water in the calorimeter to completely cover the metal sample?

2. Why did we heat the metal in a dry test tube rather than in the boiling water?

3. The water in the beaker gets its heat energy from the _____ and the water in the calorimeter gets its heat energy from the _____.

4. What is the specific heat in J/g°C for a metal sample with a mass of 95.6 g which absorbs 841 J of energy when its temperature increases from 30.0°C to 98.0°C?

5. What effect does the initial temperature of the water have on the change in temperature of the water after the hot metal is added? Explain your answer.

> Results of scientific experiments must be reproducible when repeated or they do not mean anything. When results are repeated, the experiment is said to have good **precision.** When the results agree with a theoretical value they are described as **accurate.**

6. Which is better, the precision or the accuracy of the experimental specific heats determined for your metal sample? Support your answer with your data.

> Use the data presented in Table 5.1 to answer questions 7−9 and complete the graph on the next page to show the relationship between the atomic mass and the specific heat of the seven metals listed. Make the graph following the guidelines provided in Study Aid 3.

7. a. What is the independent variable? _____

   b. What is the dependent variable? _____

8. In Table 5.1, what is the range of values for atomic mass? _____

9. In Table 5.1, what is the range of values for specific heat?_____

10. Plot the data in Table 5.1 on the graph below. Be sure to include the following:

    a. A title

    b. Placement of the independent and dependent variables on the appropriate axes

    c. Increments for each axis

    d. Labels for each axis

    e. Plotting the data points

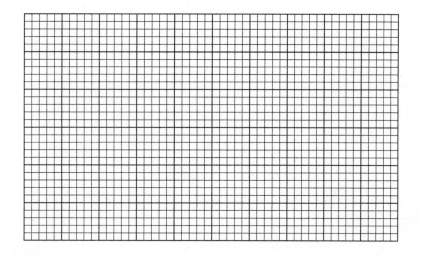

11. Use your graph to summarize the relationship between the atomic mass of metal atoms and the specific heat of a metal.

12. The specific heat was measured for two unknown metal samples. The first sample tested had a specific heat of 0.54 J/g°C. Use your graph to estimate the atomic mass of the metal.

    _____

    The second metal had a specific heat of 0.24 J/g°C. Use your graph again and estimate the atomic mass of this metal.

    _____

EXPERIMENT 6

Freezing Points — Graphing of Data

MATERIALS AND EQUIPMENT

Solids: benzoic acid (C_6H_5COOH) and crushed ice. **Liquid:** glacial acetic acid ($HC_2H_3O_2$). Thermometer, watch or clock with second hand, slotted corks or stoppers.

DISCUSSION

All pure substances, elements and compounds, possess unique physical and chemical properties. Just as one human being can be distinguished from all others by certain characteristics — fingerprints or DNA, for example — it is also possible, through knowledge of its properties, to distinguish any given compound from among the many millions that are known.

A. Melting and Freezing Points of Pure Substances

The melting point and the boiling point are easily determined physical properties that are very useful in identifying a substance. Consequently, these properties are almost always recorded when a compound is described in the chemical literature (textbooks, handbooks, journal articles, etc.). The freezing and melting of a pure substance occurs at the same temperature, measured when the liquid and solid phases of the substance are in equilibrium. When energy is being removed from a liquid in equilibrium with its solid, the process is called freezing; when energy is being added to a solid in equilibrium with its liquid, the process is called melting.

$$\text{liquid} \underset{\substack{+ \text{ energy} \\ \text{melting}}}{\overset{\substack{\text{freezing} \\ - \text{ energy}}}{\rightleftharpoons}} \text{solid}$$

In this experiment, we will determine the freezing point of a pure organic compound, glacial acetic acid ($HC_2H_3O_2$). When the experimental freezing point has been determined, it will be compared with the melting point temperature listed in the *Handbook of Chemistry and Physics.*

When heat is removed from a liquid, the liquid particles lose kinetic energy and move more slowly causing the temperature of the liquid to decrease. Finally enough heat is removed and the particles move so slowly that the liquid becomes a solid, often a crystalline solid. The temperature when this happens (the freezing point) is different for different substances.

The amount of energy removed from a quantity of liquid to freeze it, is equal to the amount of energy added to the same quantity of its solid to melt it. Thus, depending on the direction of energy flow, this equilibrium temperature is called the melting point or the freezing point.

B.   Freezing Point of Impure Substances

When a substance (solvent) is uniformly mixed with a small amount of another substance (solute), the freezing point of the resulting solution (an "impure substance") will be lower than that of the pure solvent. For example, the accepted freezing point for pure water is 0.0°C. Solutions of salt in water may freeze at temperatures as low as −21°C depending on the amount of salt added to the water. Antifreeze is added to the water in a car radiator to lower the freezing point of the water.

Melting point/freezing point data are of great value in determining the identity and/or purity of substances, especially in the field of organic chemistry. If a sample of a compound melts or freezes appreciably below the known melting point of the pure substance, we know that the sample contains impurities which have lowered the melting point. If the melting point of an unknown compound agrees with that of a known compound, the identity can often be confirmed by mixing the unknown compound with the known and determining the melting point of the mixture. If the melting point of the mixture is the same as that of the known compound, the compounds are identical. On the other hand, a lower melting point for the mixture indicates that the two compounds are not identical.

C.   Supercooling During Freezing

Frequently when a substance is being cooled, the temperature will fall below the true freezing point before crystals begin to form. This phenomenon is known as supercooling because the substance is cooled below its freezing point without forming a solid. Supercooling is more likely to occur if the liquid remains very still and undisturbed as its temperature is lowered. When the system is disturbed in any way, for example, by stirring or jarring, crystallization occurs rapidly throughout the system. As the crystals form, heat is released (called the heat of crystallization) and the temperature rises quickly to the freezing point of the substance. Thus, supercooling does not change the freezing point of the substance.

D.   Freezing Point Determinations

You will do three freezing point determinations during this experiment using the setup in Figure 6.1.

Trial 1. Freezing point determination of pure glacial acetic acid WITH STIRRING. This will usually eliminate supercooling.

Trial 2. Freezing point determination of pure glacial acetic acid WITHOUT STIRRING. This should enhance the possibility of supercooling but does not guarantee it.

Trial 3. Freezing point determination of acetic acid (the solvent) after benzoic acid (a solute) has been dissolved in it. This will be done WITHOUT STIRRING to enhance supercooling again.

The time/temperature data will be graphed and the freezing point for each trial read from the graph.

PROCEDURE

Wear protective glasses.

> **NOTES:** Since water and other contaminants will influence the freezing points in this experiment, use only clean, dry equipment.
>
> Read and record all temperatures to the nearest 0.1°C.

A. Freezing Point Determination of Pure Glacial Acetic Acid

Trial 1: With stirring

1. Fasten a utility clamp to the top of a clean, dry test tube. Position this clamp-tube assembly on a ring stand so that the bottom of the tube is about 20 cm above the ring stand base.

2. Obtain a slotted one-hole cork (or stopper) to fit the test tube (see Figure 6.1). Insert a thermometer in the cork and position it in the test tube so that the end of the bulb is about 1.5 cm from the bottom of the test tube. Turn the thermometer so that the temperature scale can be read in the slot.

3. Take your test tube, the cork/thermometer and a graduated cylinder to the fume hood. Measure out 10. mL of glacial acetic acid. Pour it into the test tube and close the test tube with the cork/thermometer. Glacial acetic acid is irritating and harmful if inhaled so keep the test tube stoppered while you work outside the hood at your bench. Rinse the graduated cylinder with water immediately.

4. Reclamp the test tube to your ring stand to minimize the risk of spilling. Make sure the thermometer bulb is covered by the acid and adjust the temperature of the acetic acid to approximately 25°C by warming or cooling the tube in a beaker of water.

5. Fill a 400 mL beaker about three-quarters full of crushed ice; add cold water until the ice is almost covered. Position the beaker of ice and water on the ring stand base under the clamped tube-thermometer assembly.

6. Read the temperature of the acetic acid and record as the 0.0 minute time reading in the Data Table. Now loosen the clamp on the ring stand and observe the second hand of your watch or clock. As the second hand crosses 12, lower the clamped tube-thermometer assembly so that all of the acetic acid in the tube is below the surface of the ice water. Fasten the clamp to hold the tube in this position.

7. Loosen the cork on the tube and stir (during Trial 1 only) the acid with the thermometer, keeping the bulb of the thermometer completely immersed in the acid. Take accurate temperature readings at 30-second intervals as the acid cools. (Zero time was when the second hand crossed 12.) Stop stirring and center the thermometer bulb in the tube as soon as you are sure that crystals are forming in the acid (one to four minutes). Circle the temperature reading when the crystals were first observed.

8. Continue to take temperature readings at 30-second intervals until a total time of 12 minutes has elapsed or until the entire volume of liquid becomes solidified. After that occurs, read the temperature for an additional 2 minutes (4 time intervals) and continue with the next step.

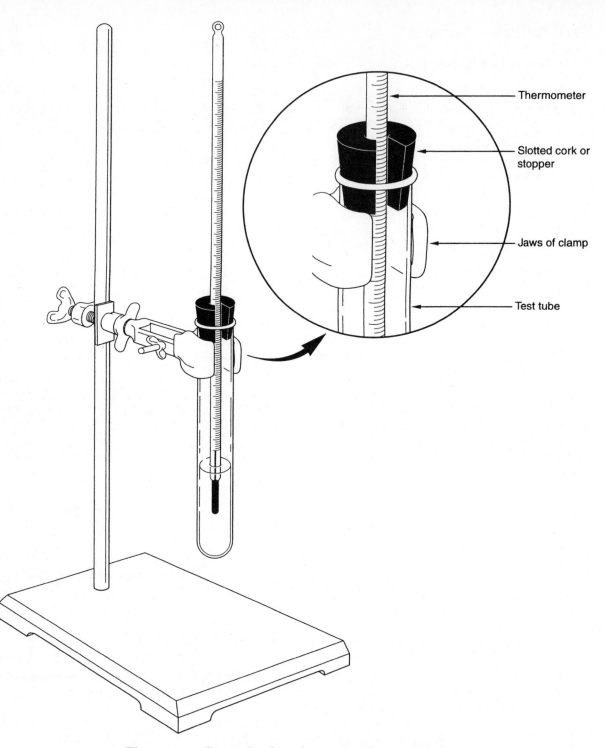

Figure 6.1   Setup for freezing-point determination

9. After completing the temperature readings, remove the test tube-thermometer assembly from the ice bath, keeping the thermometer in place. Immerse the lower portion of the test tube in a beaker of warm water to melt the frozen acetic acid. Do not discard this acid; it will be used in Trials 2 and 3.

Trial 2: Without stirring

10. Repeat steps 4–9 with the following changes:

   a. Replenish the ice bath as in step 5.

   b. After submerging the tube in the ice bath, do NOT stir. Do NOT touch or move the apparatus in any way

   c. If the temperature goes down to about 4°C or lower without the formation of acetic acid crystals and remains there, touch the thermometer and move it until crystals form which usually happens quickly. When you do this be very observant of the temperature changes. Continue to record temperature readings for the full 12 minutes or until the temperature stabilizes after crystallization for 5 minutes.

B. Freezing Point Determination of An Acetic Acid/Benzoic Acid Solution

Trial 3: Without stirring

11. Weigh approximately 0.50 g (between 0.48 and 0.52 g) of benzoic acid crystals. Now remove the thermometer from the test tube of acetic acid and lay it on the table) being careful not to contaminate the thermometer or lose any acid. Carefully add all of the benzoic acid to the acetic acid. Stir gently with the thermometer until all of the crystals have dissolved. Stir for an additional minute or two to ensure a uniform solution. Adjust the temperature of the solution to approximately 25°C.

12. Repeat step 10.

13. Warm the test tube to change the solid to a liquid and dispose of the acetic acid/benzoic acid solution in the waste container provided. Rinse the test tube with water and pour the liquid down the sink.

C. Graphing Temperature Data

Graph the three sets of data using the graph paper in the report form or prepare a computer graph. If necessary, review the instructions for preparing a graph in Study Aid 3.

REPORT FOR EXPERIMENT 6

Freezing Points–Graphing of Data

Data Table

time, minutes	Pure Acetic Acid temp, °C WITH STIRRING	Pure Acetic Acid temp, °C WITHOUT STIRRING	Impure Acetic Acid temp, °C WITHOUT STIRRING
0.0			
0.5			
1.0			
1.5			
2.0			
2.5			
3.0			
3.5			
4.0			
4.5			
5.0			
5.5			
6.0			
6.5			
7.0			
7.5			
8.0			
8.5			
9.0			
9.5			
10.0			
10.5			
11.0			
11.5			
12.0			

Graphing of Freezing Point Data

Plot your data on the graph paper or the computer using a legend as follows:

△ = Pure acetic acid with stirring

▲ = Pure acetic acid without stirring

○ = Acetic acid/benzoic acid solution without stirring

Draw rectangles around the portions of your curves that show supercooling.

QUESTIONS

Use your graph to answer the questions 1–3.

1. a. At what temperature did crystals first form in Trial 1?   _____

   b. Where did the temperature stabilize after supercooling in Trial 2?   _____

   c. What is your experimental freezing point of glacial acetic acid?   _____

   d. What is the theoretical freezing point of glacial acetic acid?   _____
      (Consult the *Handbook of Chemistry and Physics*)

2. How many degrees was the freezing point depressed by the benzoic acid?   _____

   Do this by estimating to the nearest 0.1 degree the number of degrees between the flat-test (most nearly horizontal) portions of the curves. Mark the area on the graph with an arrow (↓) to show where this temperature difference estimate was made.

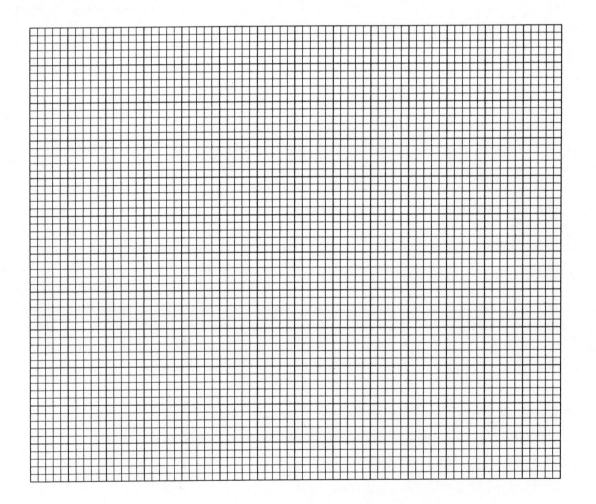

3. a. What is the effect of stirring on the freezing point of pure acetic acid?

    b. What is the effect of stirring on supercooling?

4. a. What do the melting point and freezing point of a substance have in common?

    b. What is the difference between the melting and freezing of a substance?

5. When the solid and liquid phases are in equilibrium, which phase, solid or liquid contains the greater amount of energy? Explain the rationale for your answer.

EXPERIMENT 7

Water in Hydrates

MATERIALS AND EQUIPMENT

Solids: finely ground copper(II) sulfate pentahydrate ($CuSO_4 \cdot 5H_2O$), and unknown hydrate. Cobalt chloride test paper, clay triangle, crucible and cover, 25 × 200 mm ignition test tube, watch glass.

DISCUSSION

Many salts form compounds in which a definite number of moles of water are combined with each mole of the anhydrous salt. Such compounds are called **hydrates.** The water which is chemically combined in a hydrate is referred to as **water of crystallization** or **water of hydration.** The following are representative examples:

$$CaSO_4 \cdot 2H_2O, \quad CoCl_2 \cdot 6H_2O, \quad MgSO_4 \cdot 7H_2O, \quad Na_2CO_3 \cdot 10H_2O$$

In a hydrate the water molecules are distinct parts of the compound but are joined to it by bonds that are weaker than either those forming the anhydrous salt or those forming the water molecules. In the formula of a hydrate a dot is commonly used to separate the formula of the anhydrous salt from the number of molecules of water of crystallization. For example, the formula of calcium sulfate dihydrate is written $CaSO_4 \cdot 2H_2O$ rather than $CaSO_6H_4$.

Hydrated salts can usually be converted to the anhydrous form by careful heating:

$$\text{Hydrated Salt} \xrightarrow{\Delta} \text{Anhydrous salt} + \text{water}$$

Hydrated salts can be studied qualitatively and quantitatively. In the **qualitative** part of this experiment we will observe some of properties of the liquid (water) driven off by heating the sample. In the **quantitative** part of the experiment we will determine **how much** water was in the hydrate by measuring the amount of water driven off by heating.

To make certain that all of the water in the original sample has been driven off, chemists use a technique known as **heating to constant weight.** Since time expended for this is limited, constant weight is essentially achieved when the sample is heated and weighed in successive heatings until the weight differs by no more than 0.05 g. Thus, if the second weighing is no more than 0.05 g less than the first heating, a third heating is not necessary because the sample has been heated to constant weight (almost). This is a very good reason to follow directions meticulously when heating. If the sample is not heated long enough or at the correct temperature, all of the water may not be driven off completely in the first heating.

Hence it is possible to determine the percentage of water in a hydrated salt by determining the amount of mass lost (water driven off) when a known mass of the hydrate is heated to constant weight.

$$\text{Percentage water} = \left(\frac{\text{Mass lost}}{\text{Mass of sample}} \right)(100)$$

It is possible to condense the vapor driven off the hydrate and demonstrate that it is water by testing it with anhydrous cobalt(II) chloride ($CoCl_2$). Anhydrous cobalt(II) chloride is blue but reacts with water to form the red hexahydrate, $CoCl_2 \cdot 6H_2O$.

PROCEDURE

Wear protective glasses.

A. Qualitative Determination of Water

1. Fold a 2.5 × 20 cm strip of paper lengthwise to form a V-shaped trough or chute. Load about 4 g of finely ground copper(II) sulfate pentahydrate in this trough, spreading it evenly along the length of the trough.

2. Clamp a **dry** 25 × 200 mm ignition test tube so that its mouth is 15–20 degrees **above the horizontal** (Figure 7.1a). Insert the loaded trough into the tube. Rotate the tube to a nearly vertical position (Figure 7.1b) to deposit the copper(II) sulfate in the bottom of the tube. Tap the paper chute gently if necessary, but make sure that no copper sulfate is spilled and adhering to the sides of the upper part of the tube.

3. Remove the chute and turn the tube until it slants mouth downward at an angle of 15–20 degrees **below the horizontal** (Figure 7.1c). Make sure that all of the copper(II) sulfate remains at the bottom of the tube. To obtain a sample of the liquid that will condense in the cooler part of the tube, place a clean, dry test tube, held in an upright position in either a rack or an Erlenmeyer flask, just below the mouth of the tube containing the hydrate.

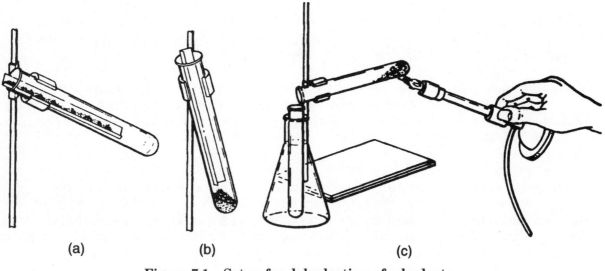

(a)   (b)   (c)

Figure 7.1   Setup for dehydration of a hydrate

4. Heat the hydrate gently at first to avoid excessive spattering. Gradually increase the rate of heating, noting any changes that occur and collecting some of the liquid that condenses in the cooler part of the tube. Continue heating until the blue color of the hydrate has disappeared, but do not heat until the residue in the tube has turned black. Finally warm the tube over its entire length—without directly applying the flame to the clamp—for a minute

or two to drive off most of the liquid that has condensed on the inner wall of the tube. Allow the tube and contents to cool.

> **NOTE:** At excessively high temperatures (above 600°C) copper(II) sulfate decomposes; sulfur trioxide is driven off and the black copper(II) oxide remains as a residue.

Observe and record the appearance and odor of the liquid that has been collected.

5. While the tube is cooling, dry a piece of cobalt chloride test paper by holding it with tongs about 20 to 25 cm above a burner flame; that is, close enough to heat but not close enough to char or ignite the paper. When properly dried, the test paper should be blue. Using a clean stirring rod, place a drop of the liquid collected from the hydrate on the dried cobalt chloride test paper. For comparison place a drop of distilled water on the cobalt chloride paper. Record your observations.

6. Empty the anhydrous salt residue in the tube onto a watch glass and divide it into two portions. Add 3 or 4 drops of the liquid collected from the hydrate to one portion and 3 or 4 drops of distilled water to the other. Compare and record the results of these tests.

 Dispose of solid residues in the waste heavy metal container provided.

B. Quantitative Determination of Water in a Hydrate

> **NOTES:**
>
> 1. **Weigh crucible and contents to the highest precision with the balance available to you.**
>
> 2. Since there is some inaccuracy in any balance, use the same balance for successive weighings of the same sample. When subtractions are made to give mass of sample and mass lost, the inaccuracy due to the balance should cancel out.
>
> 3. Handle crucibles and covers with tongs only, after initial heating.
>
> 4. Be sure crucibles are at or near room temperature when weighed.
>
> 5. **Record all data directly on the report form as soon as you obtain them.**

1. Obtain a sample of an unknown hydrate, as directed by your instructor. Be sure to record the identifying number.

2. Weigh a clean, dry crucible and cover to the highest precision of the balance.

3. Place between 2 and 3 g of the unknown into the weighed crucible. Cover and weigh the crucible and contents.

4. Place the covered crucible on a clay triangle; adjust the cover so that is slightly ajar, to allow the water vapor to escape (see Figure 7.2); and **very gently** heat the crucible for about 5 minutes. Readjust the flame so that a sharp, inner-blue cone is formed. Heat for another 12 minutes with the tip of the inner-blue cone just touching the bottom of the crucible. The crucible bottom should become dull red during this period.

5. After this first heating is completed, close the cover, cool (about 10 minutes), and weigh.

6. To determine if all the water in the sample was removed during the initial heating, reheat the covered crucible and contents for an additional 6 minutes at maximum temperature; cool and reweigh. If the sample was heated to constant weight the results of the last two weighings should agree within 0.05 g. If the decrease in mass between the two weighings is greater than 0.05 g, repeat the heating and weighing until the results of two successive weighings agree to within 0.05 g.

7. Calculate the percentage of water in your sample on the basis of the *final* weighing.

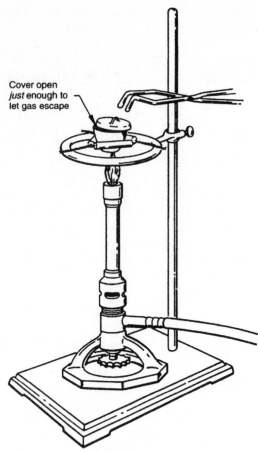

Cover open *just* enough to let gas escape

Figure 7.2   Method of heating a crucible

Dispose of the solid residue in the waste heavy metal container provided. Return the unused portion of your unknown to the instructor.

NAME _____

SECTION _____ DATE _____

REPORT FOR EXPERIMENT 7

INSTRUCTOR _____

Water in Hydrates

A. Qualitative Determination of Water

1. Describe the appearance and odor of the liquid obtained by heating copper(II) sulfate pentahydrate.

2. Compare the results observed when testing the liquid from the hydrate and distilled water with the cobalt chloride paper and the anhydrous salt by completing the table below.

Property Observed	Cobalt Chloride Paper	Anhydrous $CuSO_4$
Color before adding liquid(s) to		
Color after adding distilled water to		
Color after adding liquid from hydrate to		
Temperature change after adding distilled water	N/A	
Temperature change after adding liquid from hydrate	N/A	

B. Quantitative Determination of Water in a Hydrate

1. Mass of crucible and cover _____

2. Mass of crucible, cover, and sample _____

3. Mass of crucible, cover, and sample after 1st heating _____

4. Mass of crucible, cover, and sample after 2nd heating _____

5. Mass of crucible, cover, and sample after 3rd heating (if needed) _____

6. Mass of original sample
   Show calculation setup: _____

7. Total mass lost by sample during heating
   Show calculation setup: _____

8. Percentage water in sample   Sample No. _____
   Show calculation setup: _____

QUESTIONS AND PROBLEMS

1. What evidence did you see that indicated the liquid obtained from the copper (II) sulfate pentahydrate was water?

   (a)

   (b)

2. What was the evidence of a chemical reaction when the anhydrous salt samples were treated with the liquid obtained from the hydrate and with water?

   (a)

   (b)

3. Write a balanced chemical equation for the decomposition of copper(II) sulfate pentahydrate.

4. When the unknown was heated, could the decrease in mass have been partly due to the loss of some substance other than water? Explain.

5. A student heated a hydrated salt sample with an initial mass of 4.8702 g, After the first heating, the mass had decreased to 3.0662 g.

   (a) If the sample was heated to constant weight after reheating, what is the minimum mass that the sample can have after the second weighing? Show how you determined your answer.

   _____

   (b) The student determined that the mass lost by the sample was 1.8053. What was the percent water in the original hydrated sample? Show calculation setup.

   _____

EXPERIMENT 8

Water, Solutions, and pH

MATERIALS AND EQUIPMENT

Liquids: Decane, mineral oil, ethanol, liquid detergent. **Solids:** Dark blue ice cubes, methylene blue powder. **Solutions:** 0.020 M NaCl, 0.020 M glucose, 0.020 M I_2 in KI, 0.020 M HCl, 0.020 M NaOH, 0.020 M Na_2CO_3, 1.0 M HCl, colored water (green and red food coloring); **Equipment:** Molecular model kit, 50 mL buret, pencils (red and blue), synthetic cloth (5" × 5"), plastic rods (10"), beakers (1000 and 600 mL), microscope slides, 5 sizes glass capillary tubing (i.d. range = 0.5 − 2.75 mm), hot plates, culture dishes, metal pins (1-2"), fan, rayon fabric tubing, Pipetman (micropipetter) and tips (200μL and 1000 μL); pH 1-14 paper strips; weigh boats, pH meter, vortexer, graduated pipet (10 mL) and pipet pump.

DISCUSSION

Water is essential for life. It is the most abundant inorganic molecule in most cells, and can represent between 70% and 98% of a cell's volume. Most molecules that make up the cell are dissolved in water and the resulting mixture is called a **solution.** All the biochemical reactions necessary to sustain life occur in a solution of water. To understand cells, organisms, and life processes, it is essential to understand the physical and chemical properties of water and aqueous solutions.

A.   1. The molecular structure and polarity of water

Water molecules contain two hydrogen atoms bonded to a central oxygen atom with a chemical formula of H_2O. The bonds within the water molecule are **polar covalent.** Covalent bonds are formed when two atoms, in this case, hydrogen and oxygen, share a pair of **valence** electrons.

Covalent bonds are polar if the sharing of electrons between atoms is not equal, as between oxygen and hydrogen atoms in the water molecule. In the O-H polar covalent bond, the oxygen atom has a greater **electronegativity,** or a stronger attraction for the pair of electrons that it shares with hydrogen. The difference in electronegativities causes the oxygen atom to develop a *partial negative charge* (δ^-) and the hydrogen a *partial positive charge* δ^+, because the water molecule is not linear but has a bent, or 'V' shape. The oxygen atom at the point of the V is the *negatively* charged pole. The hydrogen atoms at the wide end of the V are the *positively* charged pole. Examine the structure of the water molecule in Figure 1.

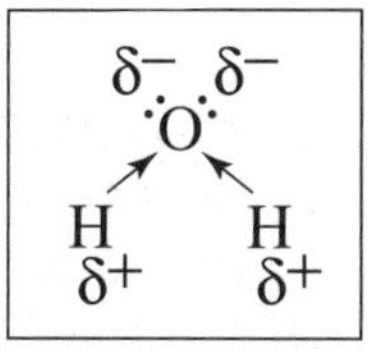

Figure 1. Polar water molecule

2. Hydrogen Bonds between Water Molecules

As a result of this polarity, water molecules are attracted to one another, with each molecule behaving like a weak magnet, each with a positive pole, and a negative pole. This attractive electrostatic force creates **hydrogen bonds** between the negative oxygen end of one water molecule and the positive hydrogen ends of another water molecule.

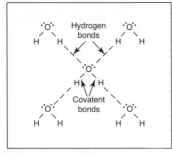

Figure 2.   Hydrogen bonding

See Figure 2. The combined force of many hydrogen bonds is responsible for many physical properties of water. It is important to remember that hydrogen bonds form between water molecules, not within a single water molecule. Hydrogen bonds between water molecules can be easily broken with the addition of kinetic or heat energy. Despite being relatively weak bonds, hydrogen bonds explain some of water's most amazing physical and chemical properties. Physical properties observed in this exercise are: Cohesion, adhesion, surface tension, specific heat, heat of vaporization, density, pH and solvent properties.

3. Cohesion and Surface Tension

Cohesion refers to the ability of any substance to stick to itself, or to stick together. Velcro is a good example of a cohesive material. The positive hydrogen poles of water molecules are attracted to the negative oxygen poles of other water molecules, resulting in cohesion. One result of cohesion is the **surface tension** of water in which water molecules are hydrogen-bonded to each other and to the water below. The collective strength of these hydrogen bonds may be strong enough to support the weight of an animal walking across the surface of a pond without breaking the surface.

4. Capillarity, Cohesion and Adhesion

Adhesion refers to the ability of any substance to stick to other things. Tape is a good example of an adhesive material. The charged portions of water molecules may also be attracted to any other oppositely charged material, resulting in adhesion. Example: Drops of water on a glass window. **Capillarity** is the ability of water to enter small pores and move up against gravity through narrow tubes. This property of water arises from an interaction between cohesion and adhesion. As water enters a narrow tube the water molecules **adhere** to the sides (interior surface area) of the tube. Each molecule will pull additional water molecules up the tube through **cohesion**.

The movement of water into the tube will stop once the upward force of adhesion is equal to the downward force of gravity acting on the volume of water. Therefore, the relationship (ratio) between the tube's interior surface area (SA) and the volume (V) of water it contains determines how high the column of water will rise. The capillarity of water, influenced by the surface area/volume ratio (SA/V) of xylem vessels, makes it possible to deliver water to the leaves at the tops of the highest trees.

5. Specific Heat and Heat of Vaporization

Water is a good insulator which means that it maintains a relatively stable temperature despite the gain or loss of heat energy (measured in **calories or joules**). **Specific heat** is defined as the amount of heat energy needed to raise the temperature of 1 gram of a substance by 1°C. A **calorie** is defined as the amount of heat energy needed to change the temperature of 1 gram of water by 1°C. Consequently, water releases a lot of heat energy as it cools and can absorb a lot of heat without increasing rapidly in temperature. This is because hydrogen bonds between water molecules stabilize their motion, and thus help keep water temperature stable. As heat energy is added to water, some heat energy is used to break hydrogen bonds. Only after the hydrogen bonds have been broken can individual water molecules move freely so that temperature (motion) can increase. (See Figure 3.)

The amount of heat needed to change 1 gram of a substance from a liquid to a vapor at its normal boiling point is the **heat of vaporization** (see Figure 3.). Hydrogen bonds must be broken in order for a water molecule to evaporate which requires energy, usually in the form of heat. As water molecules evaporate from a surface, they take energy (heat) away from that surface. Water has a high heat of vaporization because of the numerous hydrogen bonds that must be broken as water molecules evaporate.

substance	specific heat (calories/g°C)	heat of vaporization (calories/g°C)
water	1.000	540.00
diethyl ether	0.527	83.9
acetic acid	0.468	96.8
ethanol	0.456	204.0
turpentine	0.422	68.3
Benzene	0.400	94.3

Figure 3.   **Specific Heat and Heat of Vaporization**

6. Water Temperature and Density

Substances may exist in three physical states: gas, liquid, and solid. Water is the only common substance to exist in all three physical states of matter at ordinary temperatures. **Density** is a measure of the mass of a substance divided by its volume. For example, the density of liquid water at 20°C is 1.00 g/mL. For most substances, there is an inverse relationship between density and temperature. This means that the solid, or most dense phase, occurs at cooler temperatures. The gaseous, or least dense phase, occurs at higher temperatures. This relationship between temperature and density does not apply to water uniformly across the temperature range between 0°C and 100°C (the freezing and boiling points of water respectively). Between the temperatures of 4° and 0°C, the volume of water expands. Above 4°C, water behaves like other liquids and expands as it warms and hydrogen bonds are broken, contracts as it cools and more hydrogen bonds forms.

Thus, liquid water reaches its maximum density at 4°C, but there are still more hydrogen bonds that can form. As the temperature drops even lower, from 4°C to 0°C, the maximum number of hydrogen bonds is formed which requires that water molecules move apart from each other slightly. Therefore, a group of water molecules occupies a larger volume, and become less dense as it freezes. This is demonstrated when solid water, or ice, floats on liquid water. This is an important factor in allowing life to exist in ponds, and lakes.

7. Using the Density of Water to Check the Accuracy of Volume and Mass Measurements

An electronic balance that measures mass to 0.1 mg (0.0001 g) can be used to determine if an instrument measuring volume is accurate within an acceptable range which is much easier and reliable than visually assessing volume. For example, since 1 mL of water weighs 1.0 g, you can estimate the expected mass for any volume of water. For example: 1.5 mL of water should weigh 1.5 g, 0.25 mL of water should weigh 0.25 g, and 150 μL (= 0.15 mL) of water should weigh 0.15 g. The expected mass of a given volume of water is called the **theoretical mass**.

Scientists use a Pipetman micropipetter (**Figure 4.**) to measure small volumes of water and other liquids. Pipetman micropipetters dispense a very

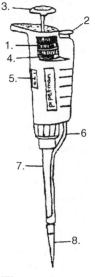

Figure 4.
Pipetman

accurate volume of liquid within a range of small measured volumes. The ranges of the micropipettors used in this experiment are L-200 (20 μL to 200 μL) and L-1000 (200 μL to 1000 μL). The size of the Pipetman is printed on the top of the plunger button (2). The micropipetter is always used with a disposable tip (8) on the end of the barrel (7). The volume is adjusted by rotating the volume adjustment dial (1) which changes the digital volume indicator (5) to display the actual volume. The plunger button (4) is depressed to eject the tip when it needs to be exchanged for a clean one. Specific directions for using the micropipetter are given in the procedure.

To determine the accuracy of volume of water dispensed by a micropipetter, the volume can be weighed on an electronic balance. By comparing the observed mass to the theoretical mass, you can make an error determination. For example, for a volume of 150.0 mL with a theoretical mass of 150.0 g and an observed mass of 148.1 g the percent error in the measurement is calculated using the following equation:

$$\frac{(\text{theoretical mass} - \text{observed (experimental) mass}) \times 100}{\text{theoretical mass}} = \% \text{ error}$$

$$\boxed{\frac{150.0 - 148.1}{150.0} \times 100 = 1.267\% \text{ error}}$$

B. Solvent properties of water

The term solution is used in chemistry to describe a system in which at least one substance is uniformly dissolved in another substance. Solutions have two components: (1) the **solvent** is the substance which does the dissolving and is the substance present in the greater quantity. In living organisms, lakes, and oceans, the solvent is water and the solution is often described as **aqueous** (*aq*). (2). the **solute(s)** is (are) the substance(s) that dissolve and remain uniformly dispersed throughout the solution after mixing. Dissolved particles of solutes are molecules and ions with a size range of 10^{-8} to 10^{-7} cm. Potential solute molecules fall into three categories: hydrophilic, hydrophobic, and amphipathic.

Hydrophilic solutes mix freely with water and include **polar** molecules and **ionic** compounds. The partially charged regions (δ^+, δ^-) of the water molecules break apart some ionic compounds like NaCl into cations like Na^+ and anions like Cl^-. These ions are then surrounded by a hydration shell of H_2O dipoles that prevent their reassociation into NaCl. The δ^+, δ^- regions of functional groups such as N-H and O-H, within polar molecules are also attracted to the oppositely charged ends of water dipoles. Thus, water disperses clumps of polar molecules such as amino acids and glucose which include polar functional groups but does not break down the molecules. This solution process enables them to move around and participate in chemical reactions

Nonpolar molecules such as hydrocarbons, fats, oils, and steroids are not soluble in water. They do not form hydrogen bonds with water and when mixed with water they form a separate insoluble layer that either rises to the surface or sinks depending on their density. These nonpolar molecules are called **hydrophobic**. Nonpolar molecules usually have a very high percent of C-H bonds and few polar functional groups. **Amphipathic** molecules such as phospholipids, have both polar and nonpolar functional groups located at different parts of the molecule. These two parts of the molecules behave differently with water.

C. 1. Solution Concentration and Molarity

The amount of solute in a given amount of solvent or solution is called the solution **concentration**. One of the most important solutes in solutions, especially those in living systems, is H^+. One source of H^+ in a solution is the covalent bonds in water itself which break spontaneously as shown in the following equation:

$$H_2O(l) \longrightarrow H^+(aq) + OH^-(aq)$$

The H^+ ion is a hydrogen atom (H) which has lost its only electron to the OH^- group, Since the H atom has only 1 proton in its nucleus and no neutrons, the H^+ ion is simply a proton. Therefore, the H^+ ion is often referred to as a proton, especially in biochemistry and cell physiology.

After a very small amount of H^+ *(aq)* forms, the covalent bond reforms and molecular water is reformed. Thus, the ionization of water is represented as an equilibrium.

$$H_2O(l) \rightleftarrows H^+(aq) + OH^-(aq)$$

The H^+ ion (proton) has a great effect on many chemical reactions, including biological processes that sustain life. Thus, it is important to know exactly how many H^+ ions (the solute) are mixed with water (the solvent) in an aqueous solution at any given time. The amount of solute in a given amount of solute or solution is called the solution **concentration.** The two most common ways to express concentration are **molarity** and **mass-percent**. In this experiment, we are working with molarity which is defined as

$$\text{Molarity (M)} = \frac{\text{Moles of solute}}{\text{liter of solution}}$$

For 1.0 L pure water, the molar concentration of H^+ has been measured as 0.0000001 mole/liter. This can also be expressed as 1×10^{-7} moles/liter and is usually converted to a more convenient unit of concentration for H^+, pH. The molarity of H^+ in pure water corresponds to a pH of 7, or neutral. Most living organisms exist in environments which are near neutral pH, although a few are adapted to pH extremes.

2. pH of solutions

The definition of pH, shown on the right, is used to convert from molarity of H^+ to pH: The H^+ written in brackets as $[H^+]$, means the concentration in moles/liter. The logarithm (log) of a number is simply the power to which 10 must be raised to give that number. Since the

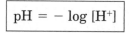
$$pH = -\log [H^+]$$

concentration of H^+ in pure water is 0.0000001 M which is also written as 1×10^{-7} M and the logarithm of $10^{-7} = -7$, then the pH = 7. (pH = $-\log (10^{-7}) = -(-7) = 7$.

For aqueous solutions the pH of the resulting solution depends on the solute being added to the water. If the solute adds to the number of H^+ ions in the mixture, the pH will decrease. Example: if 1.0 mole of HCl is added to 1 liter of water, then (assuming 100% ionization) 1.0 mole of H^+ will form in the solution. This can be expressed as 1×10^0 moles/liter, which is a pH of 0. The pH scale ranges from 0 to 14 which corresponds to molarity of 10^0 to 10^{-14} moles H^+ per liter solution. Pure water has a pH of 7. **Acids** are substances which add H^+ ions to the solution and lower the pH.

(Example: 10^{-7} M to 10^{-3} M = pH 7 to pH 3). The formula for an acid, usually starts with H (hydrogen)

$$HCl \xrightarrow{\text{H}_2\text{O}} H^+(aq) + Cl^-(aq) \qquad\qquad \text{(acid)}$$

Bases are substances which react with H^+ ions, and thus raise the pH of a solution, (Example: 10^{-7} M to 10^{-12} M = pH 7 to 12). The formula for a base usually ends in $-OH$ (hydroxide).

$$NaOH(s) \xrightarrow{\text{H}_2\text{O}} Na^+(aq) + OH^-(aq) \qquad\qquad \text{(base)}$$

Salts are substances that are neither acids nor bases, Salts are composed of oppositely charged **ions**, joined by **ionic bonds**. Ions are charged atoms or groups of atoms (such as Na^+ or Cl^-)) formed by the gain or loss of one or more electrons from the atom or group of atoms. The pH of salt solutions cannot be generalized. Some salts form neutral solutions while others form either acidic or basic solutions. For example:

$$NaCl(s) \xrightarrow{\text{H}_2\text{O}} Na^+(aq) + Cl^-(aq) \qquad\qquad \text{(salt, pH = 7.0)}$$

$$NaC_2H_3O_2(s) \xrightarrow{\text{H}_2\text{O}} Na^+(aq) + C_2H_3O_2^-(aq) \qquad \text{(salt, pH > 7.0)}$$

$$NH_4Cl(s) \xrightarrow{\text{H}_2\text{O}} NH_4^+(aq) + Cl^-(aq) \qquad\qquad \text{(salt, pH < 7.0)}$$

Buffers are solutions that minimize pH changes when small amounts of acids or bases are added to a solution.

3. Measurement of pH

To measure the pH of a solution, **acid-base indicators** such as litmus or phenolphthalein can be used. However, these indicators do not give very precise information. It is possible to get more precise hydrogen ion concentrations using **pH 1-14 paper strips** impregnated with several indicators, or with an instrument called a **pH meter**. In this experiment you will measure the pH of serial dilutions of a strong acid using both pH 1-14 paper and a pH meter. **pH meters** come in many different versions and sizes and share some of the following common features:

   a. pH meters use a glass electrode that is immersed in the solution being tested. The electrode converts the H^+ concentration into an electrical potential which is read by a voltmeter calibrated in pH units.

   b. pH meters are calibrated against standard solutions of known pH. After calibration, the electrode is immersed in the test solutions and the pH meter provides a value relative to the standard.

PROCEDURE

A. 1. Molecular Structure and Polarity of water:

   1. Use a large sphere (oxygen atom) with four pre-drilled holes from the model kit and insert one stick into each hole (four sticks total).

   2. Attach small spheres (hydrogen atoms) to two of the sticks.

   3. The sticks inserted into both the central atom (oxygen) and the peripheral hydrogen atoms (white spheres) represent a shared pair of electrons (a single covalent bond).

Each stick without a peripheral atom represents an unshared pair of electrons. The positioning of the pre-drilled holes is determined to maximize the distance between the electrons around the central atom because like charges repel each other.

4. Use the models from several students and identify where hydrogen bonds between separate water molecules will form. Draw the covalent and hydrogen bonds on the figure provided on your report form.

A. 2. Hydrogen Bonds Between Water Molecules

1. Attach a 50 mL buret filled with water to a buret stand. Place a 250 mL beaker underneath the buret to catch the water. Open the buret for a few seconds and make sure the water falls into the beaker.

2. Vigorously rub a plastic rod with a piece of soft synthetic cloth. This will cause the rod to become negatively charged.

3. Open the buret and move the charged rod slowly towards the stream of water. Position the rod so that it comes close to the stream *without becoming wet*. Observe how the stream of water is diverted by the charged rod.

4. Repeat this procedure placing the rod on the opposite side of the stream.

5. Observe how the stream of water is diverted by the charged rod and record your observations on the report form.

A. 3. Cohesion and Surface Tension

1. Obtain a clean microscope slide. Place a drop of water on one end and a drop of ethanol on the other.

2. Observe the two drops and record your observations on the report form.

3. Fill a culture dish with distilled water and wait until the water is still.

4. Using forceps gently lower a pin onto the water surface. The pin should be floating by itself. Observe the water surface under the pin.

5. Carefully add several drops of liquid detergent to the water near the pin. (DO NOT touch the pin while adding detergent!) Report your observations on the report form.

A. 4. Capillarity, Cohesion and Adhesion.

1. Submerge the tips of five pieces of glass tubing, of increasing diameter or bore size, into a dish of green water.

2. Remove the tubes from the green water and measure the height of the water column in each tube. Use these measurements to complete the report form for A4.

A. 5. Specific Heat and Heat of Vaporization

1. Obtain two 250 ml. Erlenmeyer flasks with 2-hole rubber stoppers and a thermometer inserted into one hole. Fill one flask with tap water and the other with air.

2. Place each flask in a 600 mL beaker of water on a hot plate. Turn on the hot plate to its highest setting.

3. Monitor the temperature of both flasks. As soon as each flask reaches 75°C, remove the beaker from the hot plate, and remove the flask from the beaker using hot pads or tongs. NOTE: The flasks will not reach 75°C at the same time. DO NOT pick up the flask by the thermometer or stopper. Record the temperature at 0 minutes in Data Table 2 on the report form.

4. Continue recording the temperature of each flask at 2 minute intervals for a total of 20 minutes.

5. Complete all A. 5 a-e questions on the report form, including the computer graph using Excel.

6. Secure two thermometers in a clamp supported by a ring stand. Cover the bulbs of both thermometers with a cotton sleeve. Dip one of the bulbs in a beaker of water while keeping the other dry. **Warm both bulbs with your hands until they are both at or above 30°C.**

7. Record the initial (time 0) temperature of each thermometer (wet and dry) on Data Table 3. of the report form.

8. Position a fan directly in front of both bulbs. Turn on the fan. Record the temperature of each thermometer at one minute intervals for 5 minutes.

A. 6. Water Temperature And Density

1. Fill a 1000 mL beaker with warm tap water (40-50°C) to the 800 mL mark. Place the beaker on a piece of white paper.

2. Using a 10 mL graduated pipet and a pipet pump, slowly transfer 10 mL of cold (4°C) red water to the very bottom of the beaker. Record your observations on the report form.

3. Using gloves or tongs, gently transfer a blue ice cube to the beaker. Gently float the ice cube so it does not sink or splash. Record your observations of the blue ice cube as it melts and increases in temperature from 0°C to 4°C.

4. Continue to observe the behavior of colored water in your beaker **for 5 minutes.** Record your observations and answer the questions on the report form.

A. 7. Density and Volume

1. Put a small weigh boat on the electronic balance and "tare" the balance.

2. Refer to Data Table 4 of the report form to pipet specific volumes into the weigh boat.

3. Use the L-1000 micropipet to measure the specified volumes of distilled water into the weigh boat and record the mass in the data table 4 after each volume. Tare the balance every time you pipet.

4. For each micropipet measure the specified volume of distilled water into a small weigh boat on a "tared" balance. Record the mass under observed mass.

5. Calculate the % error as described in A7 of the discussion.

6. Complete Table 4 on the report form.

B. 1. Solvent Properties of Water

1. Add 5 mL of distilled water to 8 small clean test tubes. Label the test tubes a to h.

2. Using the disposable pipets provided with each solution, add the following to each of the 8 test tubes:

   a. 2.0 mL distilled water

   b. 2.0 mL of 0.020 M NaCl (table salt)

   c. 2.0 mL of 0.020 M glucose (a sugar)

   d. 1.0 mL of 0.020 M I_2KI and 1.0 mL decane

   e. 2.0 mL of 0.020 M HCl (an acid)

   f. 2.0 mL of 0.020 M NaOH (a base)

   g. 2.0 mL of mineral oil

   h. 2.0 mL 0.020 M Na_2CO_3

3. Vigorously vortex or swirl each test tube and calculate the molarity of each solution from the information provided above. Complete the second row of Data Table 5 on the report form.

4. Save all solutions for use in B2.

B. 2. pH of Solutions

1. Use the pipet to transfer a drop of each solution in B1 to separate clean pH strips. Determine and record the pH on the report form.

2. Indicate on the report form if the solution in each test tube is an acid, a base, neutral or hydrophobic.

3. Dispose of solution B2d in the organic waste container. Dispose of all remaining solutions into the sink and flush with running water.

B 3. Measurement of pH using a pH meter

1. Set up 4 small test tubes in a rack.

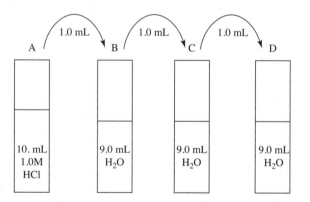

2. To test tube A, add 10. mL of 1.0 M HCl; To test tubes B, C, and D add 9.0 mL of distilled water

3. a. Use a L-1000 micropipetter to remove 1.0 mL of the 1.0 M HCl from tube A and transfer to tube B. Vortex or mix.

   b. With a fresh pipet tip, remove 1.0 mL of the solution in tube B and transfer to tube C. Vortex or mix.

c. With another fresh pipet tip, remove 1.0 mL of the solution in tube C and transfer to tube D. Vortex or mix.

4. Transfer a drop from D to a fresh piece of pH paper and record the pH on Data Table 6.

5. Continue using the same pipet tip and measure the pH of the solutions in C, B and A (in that order). Use a fresh piece of pH paper each time. Record the results on Data Table 6.

6. Go to the pH meter station and use the pH meter to measure the pH in each of the test tubes D through A (in that order). Record the results on Data Table 5.

7. Dispose of all solutions in the sink. Flush with running water.

NAME _____

SECTION _____ DATE _____

INSTRUCTOR _____

REPORT FOR EXPERIMENT 8.

Water, Solutions, and pH

Observations and Measurements

A. 1.

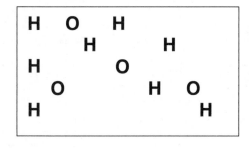

Use a red pencil to draw the covalent bonds within the molecules on the left and a blue pencil to draw the hydrogen bonds between the molecules

A. 2. Draw the stream of water falling from the buret when the rod is positioned close to the stream. Indicate a single water molecule within the stream and its orientation toward the rod. Which part of the H_2O molecule is drawn toward the rod?

A. 3. a. Draw the drops of water and alcohol on the microscope slide below. Explain the difference in the margins and height of each drop.

b. Explain why the pin floats on the surface of the water?

c. What would you predict if a bowl filled with alcohol had a pin carefully dropped on on its surface? Explain your prediction.

d. Explain why the pin sinks when detergent is added to the culture dish. Why do we add detergent to our laundry and to water used to wash dishes?

e. Predict what would happen to the pin if the temperature of the pure water in the dish were increased to 50°C. Explain your answer.

A. 4. Data Table 1: Relationship between column surface area and height of column of glass tubing

measurement	1	2	3	4	5	Average value
diameter (d), mm						$d_1 =$
height, mm						$h_1 =$
surface area, mm² $SA = \pi r^2$						$SA_1 =$
Volume, mm³ $= \pi r^2 h$						$V_1 =$
SA/V ratio mm²/mm³						$SA_1/V_1 =$

a. Complete Data Table 1. ($r = {}^1/_2$ d and $\pi = 3.14$)

b. Describe the relationship between the SA/V ratio and the height of the water column in each tube.

c. What is the role of adhesion in the movement of water in the column?

d. What is the role of cohesion in the movement of water in the column?

e. Why does the water stop rising in the column?

A. 5. Specific Heat and Heat of Vaporization

Data Table 2: Temperature of air and water as a function of time

Time (min)	0	2	4	6	8	10	12	14	16	18	20
Air flask temperature (°C)	75										
Water flask temperature (°C)	75										

   a. Use Excel (See Study Aid 3 for instructions) to make a line graph of temperature/time data in Data Table 2. Attach the graph to this lab report.

   b. Knowing that hydrogen bonding is stronger in water than in air (a gas), describe the relationship between specific heat and hydrogen bonding.

   c. Which flask absorbed more heat energy in order to reach a temperature of 75°C? Why?

   d. Both flasks have the same amount of surface area and will lose heat at the same rate when they are removed from the hot plate. Explain why the air and water flasks temperatures decrease at different rates.

   e. When you go to the beach on a sunny day you cannot walk on the sand in your bare feet because the sand is so hot. Sun transfers heat energy to both the water and the sand but the water remains cool. What can you conclude about the specific heat of sand?

Data Table 3: Temperature change over time for wet and dry bulb thermometers

Time (minutes)	0	1	2	3	4	5
Wet bulb temperature (°C)	30°C					
Dry bulb temperature (°C)	30°C					

   f. The vaporization of water requires the addition of heat energy (calories). Think about the source of the heat that is absorbed by water as it evaporates. Explain why the wet thermometer cools more quickly mat the dry thermometer.

g. Mammals are endothermic animals that maintain a constant body temperature. The ability to maintain a consistent body temperature is an example of homeostasis. Explain how a human body reduces its temperature on a hot day, or during vigorous exercise to maintain homeostasis.

A. 6. Water Temperature and Density

a. Describe what happened when 10. mL of red water with a temperature of 4°C was added to a beaker of room temperature water.

b. Explain why the red water mixed with the room temperature water over time.

c. Explain why the blue ice cubes floated and the blue solution gradually mixed with the room temperature water.

d. What does it mean when the color of the water is uniform in this exercise? (one sentence)

A. 7. Density and Volume

Data Table 4: Pipeting precision: mass vs. volume

Micropipette	Volume, μL	Volume, mL	Expected mass, g	Observed mass, g	% Error
L-1000					
L-1000					
L-1000					
L-200					
L-200					
L-200					

B. 1. Solvent Properties of Water

Data Table 5:

	a. H_2O	b. NaCl (aq)	c. Glucose (aq)	d. Decane	e. HCl (aq)	f. NaOH (aq)	g. Mineral Oil	h. Na_2CO_3 (aq)
calculated Molarity	N/A			N/A			N/A	
calculated pH	N/A	N/A	N/A	N/A			N/A	NA
Indicator measured pH				N/A			N/A	
Acid/base or neutral				N/A			N/A	

Use the formula $M_1V_1 = M_2V_2$ and calculate the molarity of the solutions after the dilution of the original 0.020 M solutions with water in each test tube. Show the calculation set-up.

B. 2. pH of Solutions

a. Convert the molarity of HCl and NaOH to pH. Show your calculation set-up.

b. Record the pH measured by the indicator paper and complete the rest of Data Table 5.

c. Which of the above solutes are hydrophobic?

d. Is the iodine solution more soluble in water or in decane? How do you know?

e. Which solvent is more dense, decane or water? What experimental evidence supports your answer?

B. 3. Measurement of pH Using a pH meter

Data Table 6: Measurement of pH in serial dilutions of 1 M HCl

Test tube	A	B	C	D
Molarity, M	1.0			
M, 10^x				
pH (paper)				
pH (meter)				

QUESTIONS AND PROBLEMS

1. In plants, xylem vessels transport water from the roots, where it is absorbed, up towards the leaves, where it is consumed. Capillarity is partly responsible for this phenomenon. Use the average values in Data Table 1 and the equation: $d_1h_1 = d_2h_2$ to solve the following problem.

   a. How high (h_2) will water rise in a xylem vessel that has a diameter (d_2) of 5.0 μm? Convert μm to **mm** using the conversion factor: 1000 μm = 1 mm. Use the average values of your lab data for d_1 and h_1. Show your calculation setups:

   Conversion of 5.0 μm to mm

   Height of water column in xylem vessel

   _____ mm= _____ m

   b. What is the volume of water (in mm³) that is contained within this xylem vessel? Show your calculation setup:

   _____ mm³

   c. 1.0 cm³ of water weighs 1.0 gram. What is the weight of the water inside the xylem vessel? (hint: first calculate the volume of water in the xylem vessel, then use the conversion factor: 1cm³ = 100 mm³). Show your calculations set up:

   _____ g

EXPERIMENT 9

Properties of Solutions

MATERIALS AND EQUIPMENT

Solids: ammonium chloride (NH_4Cl), barium chloride ($BaCl_2$), barium sulfate ($BaSO_4$), fine and coarse crystals of sodium chloride ($NaCl$), and sodium sulfate (Na_2SO_4). **Liquids:** decane ($C_{10}H_{22}$), isopropyl alcohol (C_3H_7OH), and kerosene. **Solutions:** saturated iodine-water (I_2), and saturated potassium chloride (KCl).

DISCUSSION

Solute, Solvent, and Solution

The term **solution** is used in chemistry to describe a homogeneous mixture in which at least one substance (the **solute**) is dissolved in another substance (the **solvent**). The solvent is the substance present in greater quantity and the name of the solution is taken from the name of the solute. Thus, when sodium chloride is dissolved in water, sodium chloride is the solute, water is the solvent, and the solution is called a sodium chloride solution.

In this experiment we will be working with two common types of solutions: those in which a solid solute is dissolved in a liquid solvent (water), and a few in which a liquid solute is dissolved in a liquid solvent.

Like other mixtures, a solution has variable composition, since more or less solute can be dissolved in a given quantity of a solvent. The amount of solute that remains uniformly dispersed throughout the solution after mixing is referred to as the **solution concentration** and can be expressed in many different ways. The maximum concentration that a solution can have varies depending on many factors, including the temperature, the kind of particles in the solute, and interactions between the solute particles and the solvent. In general, water, which is polar, is a better solvent for inorganic than for organic substances. On the other hand, nonpolar solvents such as benzene, decane, and ether are good solvents for many organic substances that are practically insoluble in water.

Dissolved solute particles can be either molecules or ions and their size is of the order of 10^{-8} to 10^{-7} cm (1-10 Å). Many substances will react chemically with each other only when they are dissociated into ions in solution. For example, when the two solids sodium chloride ($NaCl$) and silver nitrate ($AgNO_3$) are mixed, no detectable reaction is observed. However, when aqueous solutions of these salts are mixed, their component ions react immediately to form a white precipitate ($AgCl$).

The rate at which a solute and solvent will form a solution depends on several factors, all of which are related to the amount of contact between the solute particles and the solvent. A solid can dissolve only at the surface that is in contact with the solvent. Any change which increases that contact will increase the rate of solution and vice versa. Thus, the rate of dissolving a solid solute depends on:

1. The particle size of the solute
2. Agitation or stirring of the solution
3. The temperature of the solution
4. The concentration of the solute in solution

Solubility, Miscibility, and Concentration

The term **solubility** refers to the maximum amount of solute that will dissolve in a specified amount of solvent under stated conditions. At a specific temperature, there is a limit to the amount of solute that will dissolve in a given amount of solvent.

Solubility can be expressed in a relative, qualitative way. For example a solute may be very soluble, moderately soluble, slightly soluble, or insoluble in a given solvent at a given temperature. Table 8.1 shows how temperature effects the amount of four different salts that dissolve in 100 g of water.

Table 9.1
Temperature Effect on Solubility of Four Salts in Water, g solute/100 g water

	0°C	10°C	20°C	30°C	40°C	50°C	60°C	70°C	80°C	90°C	100°C
KCl	27.6	31.0	34.0	37.0	40.0	42.6	45.5	48.3	51.1	54.0	55.6
NaCl	35.7	35.8	36.0	36.3	36.6	37.0	37.3	37.8	38.4	39.0	39.8
KBr	53.5	59.5	65.2	70.6	75.5	80.2	85.5	90.0	95.0	99.2	104.0
BaCl$_2$	31.6	33.3	35.7	38.2	40.7	43.6	46.6	49.4	52.6	55.7	58.8

The term **miscibility** describes the solubility of two liquids in each other. When both the solute and solvent are liquids, their solubility in each other is described as miscible (soluble) or immiscible (insoluble). For example, ethyl alcohol and water are miscible; oil and water are immiscible.

The **concentration** of a solution expresses how much solute is dissolved in solution and can be expressed several ways:

1. **Dilute vs. Concentrated:** a dilute solution contains a relatively small amount of solute in a given volume of solution; a concentrated solution contains a relatively large amount of solute per unit volume of solution.

2. **Saturated vs. Unsaturated vs. Supersaturated:**

   a. A **saturated** solution contains as much dissolved solute as possible at a given temperature and pressure. The dissolved solute is in equilibrium with undissolved solute. A saturated solution can be dilute or concentrated. The solutions described in Table 8.1 are saturated at each temperature.

   Solute(solid) $\rightleftharpoons$ Solute(dissolved)

b. **Unsaturated** solutions contain less solute per unit volume than the corresponding saturated solution. Thus, more solute will dissolve in an unsaturated solution (until saturation is reached).

c. **Supersaturated** solutions contain more dissolved solute than is normally present in the corresponding saturated solution. However, a supersaturated solution is in a very unstable state and will form a saturated solution if disturbed. For example, when a small crystal of the dissolved salt is dropped into a supersaturated solution, crystallization begins at once and salt precipitates until a saturated solution is formed.

3. **Mass-percent Solution** is a quantitative expression of concentration expressed as the percent by mass of the solute in a solution. For example, a 10% sodium hydroxide solution contains 10 g of NaOH in 100 g of solution (10 g NaOH + 90 g H_2O); 2 g NaOH in 20 g of solution (2 g NaOH + 18 g H_2O). The formula for calculating mass percent is:

$$\text{Mass percent} = \left(\frac{\text{g solute}}{\text{g solute} + \text{g solvent}}\right)(100)$$

4. **Mass per 100 g solvent** is another quantitative expression of concentration (and the one used in Table 8.1). It is not the same as the Mass percent concentration above because the units are g solute/100 g solvent. Thus, for the 10% NaOH solution described in No. 3, the g NaOH/100 g H_2O would be calculated as follows:

$$\left(\frac{10 \, \text{g NaOH}}{90 \, \text{g H}_2\text{O}}\right)(100) = \frac{11 \, \text{g NaOH}}{100 \, \text{g H}_2\text{O}}$$

5. **Molarity** is the most common quantitative expression of concentration. Molarity is the number of moles (molar mass) of solute per liter of solution. Thus a solution containing 1 mole of NaOH (40.00 g) per liter is 1 molar (abbreviated 1 M). The concentration of a solution containing 0.5 mole in 500 mL (0.5 L) is also 1 M. The formula for calculating molarity is:

$$\text{Molarity} = \frac{\text{moles of solute}}{\text{liter of solution}} = \frac{\text{moles}}{\text{liter}}$$

PROCEDURE

Wear protective glasses.

A. Concentration of a Saturated Solution

Use the same balance for all weighings.
Make all weighings to the highest precision of the balance.

1. Prepare a water bath with a 400 mL beaker half full of tap water and heat to boiling. (See Figure 1.6.)

2. Weigh an empty evaporating dish. Obtain 6 mL of saturated potassium chloride solution and pour it into the dish. Weigh the dish with the solution in it and record these masses on the report form.

3. Place the evaporating dish on the beaker of boiling water and continue to boil until the potassium chloride solution has evaporated almost to dryness (about 25 to 30 minutes), **adding more water to the beaker as needed.**

> While the evaporation is proceeding, continue with other parts of the experiment.

4. Remove the evaporating dish and beaker from the wire gauze and dry the bottom of the dish with a towel. Put the dish on the wire gauze and heat gently for 1-2 minutes to evaporate the last traces of water. Do not heat too strongly because at high temperatures there is danger of sample loss by spattering.

5. Allow the dish with dry potassium chloride to cool on the Ceramfab pad for 5 to 10 minutes and weigh. To be sure that all the water has evaporated from the potassium chloride, put the dish back on the wire gauze and heat gently again for 1-2 minutes.

6. Allow the dish to cool again on the Ceramfab pad for 5 to 10 minutes and reweigh. The second weighing should be no more than 0.05 g less than the first weighing. Otherwise a third heating and weighng should be done.

7. Add water to the residue in the dish to redissolve the potassium chloride. Pour the solution into the sink and flush generously with water.

B. Relative Solubility of a Solute in Two Solvents

1. Add about 2 mL of decane and 5 mL of water to a test tube, stopper it, and shake gently for about 5 seconds. Allow the liquid layers to separate and note which liquid has the greatest density.

2. Now, add 5 mL of saturated iodine-water to the test tube, note the color of each layer, insert the stopper, and shake gently for about 20 seconds. Allow the liquids to separate and again note the color of each layer.

3. Dispose of the mixture in this test tube in the bottle labeled **Decane Waste.**

C. Miscibility of Liquids

1. Take three dry test tubes and add liquids to each as follows:

    a. 1 mL kerosene and 1 mL isopropyl alcohol

    b. 1 mL kerosene and 1 mL water

    c. 1 mL water and 1 mL isopropyl alcohol

2. Stopper each tube and mix by shaking for about 5 seconds. Note which pairs are miscible. Dispose of the kerosene mixtures (a and b) in the bottle labeled **Kerosene Waste.** Dispose the contents in test tube (c) in the sink.

D. Effect of Particle Size on Rate of Dissolving

1. Fill a dry test tube to a depth of about 0.5 cm with fine crystals of sodium chloride. Fill another dry tube to the same depth with coarse sodium chloride crystals. Add 10 mL of tap water to each tube and stopper. Shake both tubes at the same time, noting the number of seconds required to dissolve the salt in each tube. (Don't shake the tubes for more than two minutes.)

 2. Dispose of these solutions in the sink.

E. Effect of Temperature on Rate of Dissolving

1. Weigh two 0.5 g samples of fine sodium chloride crystals.

2. Take a 100 mL and a 150 mL beaker and add 50 mL tap water to each. Heat the water in the 150 mL beaker to boiling and allow it to cool for about 1 minute.

3. Add the 0.5 g samples of salt to each beaker and observe the time necessary for the crystals to dissolve in the hot water (do not stir).

4. As soon as the crystals are dissolved in the hot water, take the beaker containing the hot solution in your hand, slowly tilt it back and forth, and observe the layer of denser salt solution on the bottom. Repeat with the cold-water solution.

 5. Dispose of these solutions in the sink.

F. Solubility versus Temperature; Saturated and Unsaturated Solutions

1. Label four weighing boats or papers as follows and weigh the stated amounts onto each one.

   a. 1.0 g NaCl   b. 1.4 g NaCl   c. 1.0 g NH_4Cl d. 1.4 g NH_4Cl

2. Record observations in the table provided on the report form as you proceed through 3-6.

3. Add the 1.0 g samples of NaCl and NH_4Cl to separate tubes labeled A and B as shown. Add 5 mL of distilled water to each, stopper and shake until each salt is dissolved.

4. Now add 1.4 g NaCl to test tube A. Add 1.4 g NH_4Cl to test tube B. Stopper and shake for about 3 minutes. Note whether all of the crystals have dissolved.

5. Place both tubes (unstoppered) into a beaker of boiling water, shake occasionally, and note the results after about 5 minutes.

6. Remove the tubes and cool in running tap water for about 1 minute. Let stand for a few minutes and record what you observe.

 7. Dispose of these solutions in the sink. Flush generously with water.

G. Ionic Reactions in Solution

1. Into four labeled test tubes, place pea-sized quantities of the following salts, one salt in each tube: (a) barium chloride, (b) sodium sulfate, (c) sodium chloride, (d) barium sulfate.

2. Add 5 mL of water to each tube, stopper, and shake to dissolve. One of the four salts does not dissolve.

3. Mix the barium chloride and sodium sulfate solutions together. Note the results. (Sodium chloride and barium sulfate are the products of this reaction.)

 Dispose of all tubes containing barium in the waste bottle provided. The remaining tubes can be rinsed in the sink.

REPORT FOR EXPERIMENT 9

Properties of Solutions

A.   Concentration of Saturated Solution

1. Mass of empty evaporating dish _____

2. Mass of dish + saturated potassium chloride solution _____

3. Mass of dish + dry potassium chloride, 1st heating _____

4. Mass of dish + dry potassium chloride, 2nd heating _____

5. Mass of saturated potassium chloride solution _____
   Show Calculation Setup

6. Mass of potassium chloride in the saturated solution _____
   Show Calculation Setup

7. Mass of water in the saturated potassium chloride solution _____
   Show Calculation Setup

8. Mass percent of potassium chloride in the saturated solution _____
   Show Calculation Setup

9. Grams of potassium chloride per 100 g of water (experimental) _____
   in the original solution.
   Show Calculation Setup

10. Grams of potassium chloride per 100 g of water (theoretical) _____
    (From Table 8.1) at 20°C.

B.   Relative Solubility of a Solute in Two Solvents

1. (a) Which liquid is denser, decane or water?   _____

   (b) What experimental evidence supports your answer?

2. Color of iodine in water: _____

   Color of iodine in decane: _____

3. (a) In which of the two solvents used is iodine more soluble?   _____

   (b) Cite experimental evidence for your answer.

C.   Miscibility of Liquids

1. Which liquid pairs tested are miscible?

2. How do you classify the liquid pair decane—H_2O, miscible or immiscible?

   _____

D.   Rate of Dissolving Versus Particle Size

1. Time required for fine salt crystals to dissolve   _____

2. Time required for coarse salt crystals to dissolve   _____

3. Since the amount of salt, the volume of water, and the temperature of the systems were identical in both test tubes, how do you explain the difference in time for dissolving the fine vs. the coarse salt crystals?

E.   Rate of Dissolving Versus Temperature

1. Under which condition, hot or cold, did the salt dissolve faster?   _____

2. Since the amount of salt, the volume of water, and the texture of the salt crystals were identical in both best tubes, how do you explain the difference in time for dissolving at the hot vs. cold temperatures?

F. Solubility vs. Temperature; Saturated and Unsaturated Solutions

    Data Table: Circle the choices which best describe your observations.

	NaCl	**NH₄Cl**
1.0 g + 5 mL water	dissolved completely? yes/no saturated or unsaturated?	dissolved completely? yes/no saturated or unsaturated?
1.0 g + 5 mL water + 1.4 g	dissolved completely? yes/no saturated or unsaturated?	dissolved completely? yes/no saturated or unsaturated?
2.4 g + 5 mL water + heat	dissolved completely? yes/no saturated or unsaturated?	dissolved completely? yes/no saturated or unsaturated?
2.4 g + 5 mL water after cooling	dissolved completely? yes/no saturated or unsaturated?	dissolved completely? yes/no saturated or unsaturated?

G. Ionic Reactions in Solution

    1. Write the word and formula equations representing the chemical reaction that occurred between the barium chloride solution, $BaCl_2(aq)$, and the sodium sulfate solution, $Na_2SO_4(aq)$.

      Word Equation:

      Formula Equation:

    2. (a) Which of the products is the white precipitate?   _____

      (b) What experimental evidence leads you to this conclusion?

SUPPLEMENTARY QUESTIONS AND PROBLEMS

    1. Use the solubility data in Table 9.1 to answer the following:
      Show Calculations

      (a) What is the percentage by mass of NaCl in a saturated solution of sodium chloride at 50°C?

                                          _____

(b) Calculate the solubility of potassium bromide at 23°C. Hint: Assume that the solubility increases by an equal amount for each degree between 20°C and 30°C.

_____

(c) A saturated solution of barium chloride at 30°C contains 150 g water. How much additional barium chloride can be dissolved by heating this solution to 60°C?

_____

2. A solution of KCl is saturated at 50°C.
   Use Table 9.1

   (a) How many grams of solute are dissolved in 100 g of water?   _____

   (b) What is the total mass of the solution?   _____

   (c) What is the mass percent of this solution at 50°C?   _____

   (d) If the solution is heated to 100°C, how much more KCl can be dissolved in the solution without adding more water?

   _____

   (e) If the solution is saturated at 100°C and then cooled to 30°C, how many grams of solute will precipitate out?

   _____

EXPERIMENT 10

Composition of Potassium Chlorate

MATERIALS AND EQUIPMENT

Solids: Reagent Grade potassium chlorate ($KClO_3$) and potassium chloride (KCl). **Solutions:** dilute (6 M) nitric acid (HNO_3) and 0.1 M silver nitrate ($AgNO_3$). Two No. 0 crucibles with covers; Ceramfab pad.

DISCUSSION

The **percentage composition** of a compound is the percentage by mass of each element in the compound. If the formula of a compound is known, the percentage composition can be calculated from the molar mass and the total mass of each element in the compound. The **molar mass** of a compound is determined by adding up the atomic masses of all the atoms making up the formula. The **total mass** of an element in a compound is determined by multiplying the atomic mass of that element by the number of atoms of that element in the formula. The percentage of each element is then calculated by dividing its total mass in the compound by the molar mass of the compound and multiplying by 100.

The percentage composition of many compounds may be directly determined or verified by experimental methods. In this experiment the percentage composition of potassium chlorate will be determined both experimentally and from the formula.

When potassium chlorate is heated to high temperatures (above 400°C) it decomposes to potassium chloride and elemental oxygen, according to the following equation:

$$2\,KClO_3(s) \xrightarrow{\Delta} 2\,KCl(s) + 3\,O_2(g)$$

The relative amounts of oxygen and potassium chloride are measured by heating a weighed sample of potassium chlorate until all of the oxygen has been released from the sample. This is accomplished when the sample is heated to constant weight. In this experiment you will heat, cool, and weigh the sample at least twice. If the sample loses more than 0.05 g after the second heating it has not been heated to constant weight and should be heated a third time.

From the experiment we obtain the following three values:

1. Mass of original sample ($KClO_3$).

2. Mass lost when sample was heated (Oxygen).

3. Mass of residue (KCl).

From these experimental values (and a table of atomic masses) we can calculate the following:

4. Percentage oxygen in sample (Experimental value)

$$= \left(\frac{\text{Mass lost by sample}}{\text{Original sample mass}} \right)(100)$$

5. Percentage KCl in sample (Experimental value)

$$= \left(\frac{\text{Mass of residue}}{\text{Original sample mass}} \right)(100)$$

6. Percentage oxygen in $KClO_3$ from formula (Theoretical value)

$$= \left(\frac{3 \, \text{at. masses of oxygen}}{\text{Molar mass of } KClO_3} \right)(100) = \left(\frac{3 \times 16.00 \, g}{122.6 \, g} \right)(100)$$

7. Percentage KCl in $KClO_3$ from formula (Theoretical value)

$$= \left(\frac{\text{Molar mass of KCl}}{\text{Molar mass of } KClO_3} \right)(100) = \left(\frac{74.55 \, g}{122.6 \, g} \right)(100)$$

8. Percentage error in experimental oxygen determination

$$= \left(\frac{\text{Theoretical value} - \text{Experimental value}}{\text{Theoretical value}} \right)(100)$$

PROCEDURE

 PRECAUTIONS: Since potassium chlorate is a strong oxidizing agent it may cause fires or explosions if mixed or heated with combustible (oxidizable) materials such as paper. Observe the following safety precautions when working with potassium chlorate:

1 **Wear protective glasses.**

2. Use clean crucibles that have been heated and cooled prior to adding potassium chlorate.

3. Use Reagent Grade potassium chlorate.

 4. **Dispose of any excess or spilled potassium chlorate as directed by your instructor. (Potassium chlorate may start fires if mixed with paper or other solid wastes.)**

5. Heat samples slowly and carefully to avoid spattering molten material—and to avoid poor experimental results.

NOTES:

1. Make all weighings to the highest precision possible with the balance available to you. Use the same balance to make all weighings for a given sample. Record all data directly on the report sheet as they are obtained.

2. Duplicate samples of potassium chlorate are to be analyzed, if two crucibles are available.

3. For utmost precision, handle crucibles with tongs after the initial heating.

A. Determining Percentage Composition

Place a clean, dry crucible (uncovered) on a clay triangle and heat for 2 or 3 minutes at the maximum flame temperature. The tip of the sharply defined inner-blue cone of the flame should almost touch and heat the crucible bottom to redness. Allow the crucible to cool. If two crucibles are being used, carefully transfer the first to a Ceramfab pad and heat the second while the first crucible is cooling.

Weigh the cooled crucible and its cover; add between 1 and 1.5 g of potassium chlorate; weigh again.

> **NOTE:** The crucible must be covered when potassium chlorate is being heated in it.

Place the covered crucible on the clay triangle and **heat gently for 8 minutes** with the tip of the inner-blue cone of the flame 6 to 8 cm (about 2.5 to 3 in.) below the crucible bottom. Then carefully lower the crucible or raise the burner until the tip of the sharply defined inner-blue cone just touches the bottom of the crucible, and heat for an additional 10 minutes. The bottom of the crucible should be heated to a dull red color during this period.

Grasp the crucible just below the cover with the concave part of the tongs and very carefully transfer it to a Ceramfab pad. Allow to cool (about 10 minutes) and weigh. Begin analysis of a second sample while the first is cooling.

After weighing, reheat the first sample for an additional 6 minutes at the maximum flame temperature (bottom of the crucible heated to a dull red color); cool and reweigh. If the residue is at constant weight, the last two weighings should be in agreement. If the mass decreased more than 0.05 g between these two weighings, repeat the heating and weighing until two successive weighings agree within 0.05 g. Use the final weight in your calculations.

Complete the analysis of the second sample following the same procedure used for the first.

B. Qualitative Examination of Residue

This part of the experiment should be started as soon as the final heating and weighing of the first sample is completed and while the second sample is in progress.

Number and place three clean test tubes in a rack. Put a pea-sized quantity of potassium chloride into tube No. 1 and a like amount of potassium chlorate into tube No. 2. Add 10 mL of distilled water to each of these two tubes and shake to dissolve the salts. Now add distilled water to the crucible containing the residue from the first sample so it is one-half full. Heat the uncovered crucible very gently for about 1 minute; transfer 1 to 2 mL of the resulting solution from the crucible to tube No. 3; add about 10 mL of distilled water and mix.

Test the solution in each tube as follows: Add 5 drops of dilute (6 M) nitric acid and 5 drops of 0.1 M silver nitrate solution. Mix thoroughly. Record your observations. This procedure using nitric acid and silver nitrate is a general test for chloride ions. The formation of a white precipitation is a positive test and indicates the presence of chloride ions. A positive test is obtained with any substance that produces chloride ions in solution.

 Dispose of solutions and precipitates containing silver in the heavy metal waste container provided. Dispose of the remaining contents in the crucible down the sink.

REPORT FOR EXPERIMENT 10　INSTRUCTOR _____

Composition of Potassium Chlorate

A.　Determining Percentage Composition	Sample 1	Sample 2
1.　Mass of crucible + cover	_____	_____
2.　Mass of crucible + cover + sample before heating	_____	_____
3.　Mass of crucible + cover + residue after 1st heating	_____	_____
4.　Mass of crucible + cover + residue after 2nd heating	_____	_____
5.　Mass of crucible + cover + residue after 3rd heating (if necessary)	_____	_____
6.　Mass of original sample Show sample 1 calculation setup:	_____	_____
7.　Mass lost (total) during heating Show sample 1 calculation setup:	_____	_____
8.　Final mass of residue Show sample 1 calculation setup:	_____	_____
9.　Experimental percent oxygen in sample ($KClO_3$) Show sample 1 calculation setup:	_____	_____
10.　Experimental percent KCl in sample ($KClO_3$) Show sample 1 calculation setup:	_____	_____
11.　Theoretical percent oxygen in $KClO_3$ Show calculation setup:	_____	
12.　Theoretical percent KCl in $KClO_3$ Show calculation setup	_____	
13.　Percent error in experimental % oxygen determination Show sample 1 calculation setup	_____	_____

B. Qualitative Examination of Residue

1. Record what you observed when silver nitrate was added to the following:

(a) Potassium chloride solution

(b) Potassium chlorate solution

(c) Residue solution

2. (a) What evidence did you observe that would lead you to believe that the residue was potassium chloride?

(b) What would happen if you added silver nitrate to a solution of sodium chloride? Explain your answer.

(c) Did the evidence obtained in the silver nitrate tests of the three solutions prove conclusively that the residue actually was potassium chloride? Explain?

QUESTIONS AND PROBLEMS

1. A student forgot to read the label on the jar carefully and put potassium chloride in the crucible instead of potassium chlorate. How would the results turn out?

2. What if a potassium chlorate sample is contaminated with KCl. Would the experimental % oxygen be higher or lower than the theoretical % oxygen? Explain your answer.

3. What if a potassium chlorate sample is contaminated with moisture. Would an analysis show the experimental % oxygen higher or lower than the theoretical % oxygen? Explain your answer?

4. Calculate the percentage of Cl in $Al(ClO_3)_3$ _____

5. Other metal chlorates when heated show behavior similar to that of potassium chlorate yielding metal chlorides and oxygen. Write the balanced formula equation for the reaction to be expected when calcium chlorate, $Ca(ClO_3)_2$ is heated.

EXPERIMENT 11

Double Displacement Reactions

MATERIALS AND EQUIPMENT

Solid: sodium sulfite (Na_2SO_3). **Solutions:** dilute (6 M) ammonium hydroxide (NH_4OH), 0.1 M ammonium chloride (NH_4Cl), 0.1 M barium chloride ($BaCl_2$), 0.1 M calcium chloride ($CaCl_2$), 0.1 M copper(II) sulfate ($CuSO_4$), dilute (6 M) hydrochloric acid (HCl), concentrated (12 M) hydrochloric acid (HCl), 0.1 M iron(III) chloride ($FeCl_3$), dilute (6 M) nitric acid (HNO_3), 0.1 M potassium nitrate (KNO_3), 0.1 M silver nitrate ($AgNO_3$), 0.1 M sodium carbonate (Na_2CO_3), 0.1 M sodium chloride (NaCl), 10 percent sodium hydroxide (NaOH), dilute (3 M) sulfuric acid (H_2SO_4), and 0.1 M zinc nitrate [$Zn(NO_3)_2$]. Medicine dropper.

DISCUSSION

Double displacement reactions are among the most common of the simple chemical reactions and are comparatively easy to study.

In each part of this experiment two aqueous solutions, each containing positive and negative ions, will be mixed in a test tube. Consider the hypothetical reaction.

$$AB + CD \longrightarrow AD + CB$$

where AB(aq) exists as A^+ and B^- ions in solution and CD(aq) exists as C^+ and D^- ions in solution. As the ions come in contact with each other, there are six possible combinations that might conceivably cause chemical reaction. Two of these combinations are the meeting of ions of like charge; that is, $A^+ + C^+$ and $B^- + D^-$. But since like charges repel, no reaction will occur. Two other possible combinations are those of the original two compounds; that is, $A^+ + B^-$ and $C^+ + D^-$. Since we originally had a solution containing each of these pairs of ions, they can mutually exist in the same solution; therefore they do not recombine. Thus the two possibilities for chemical reaction are the combination of each of the positive ions with the negative ion of the other compound; that is, $A^+ + D^-$ and $C^+ + B^-$. Let us look at some examples.

Example 1. When solutions of sodium chloride and potassium nitrate are mixed, the equation for the double displacement reaction (hypothetical) is

$$NaCl(aq) + KNO_3(aq) \longrightarrow KCl(aq) + NaNO_3(aq)$$

We get the hypothetical products by simply combining each positive ion with the other negative ion. But has there been a reaction? When we do the experiment, we see no evidence of reaction. There is no precipitate formed, no gas evolved, and no obvious temperature change. Thus we must conclude that no reaction occurred. Both hypothetical products are soluble salts, so the ions are still present in solution. We can say that we simply have a solution of four kinds of ions, Na^+, Cl^-, K^+, and NO_3^-.

The situation is best expressed by changing the equation to

$$NaCl(aq) + KNO_3(aq) \longrightarrow No\ reaction$$

Example 2. When solutions of sodium chloride and silver nitrate are mixed, the equation for the double displacement reaction (hypothetical) is

$$NaCl + AgNO_3 \longrightarrow NaNO_3 + AgCl$$

A white precipitate is produced when these solutions are mixed. This precipitate is definite evidence of a chemical reaction. One of the two products, sodium nitrate ($NaNO_3$) or silver chloride ($AgCl$), is insoluble. Although the precipitate can be identified by further chemical testing, we can instead look at the **Solubility Table in Appendix 5** to find that sodium nitrate is soluble but silver chloride is insoluble. We may then conclude that the precipitate is silver chloride and indicate this in the equation with an (s). Thus

$$NaCl(aq) + AgNO_3(aq) \longrightarrow NaNO_3(aq) + AgCl(s)$$

Example 3. When solutions of sodium carbonate and hydrochloric acid are mixed, the equation for the double displacement reaction (hypothetical) is

$$Na_2CO_3(aq) + 2\,HCl(aq) \longrightarrow 2\,NaCl(aq) + H_2CO_3(aq)$$

Bubbles of a colorless gas are evolved when these solutions are mixed. Although this gas is evidence of a chemical reaction, neither of the indicated products is a gas. But carbonic acid, H_2CO_3, is an unstable compound and readily decomposes into carbon dioxide and water.

$$H_2CO_3(aq) \longrightarrow H_2O(l) + CO_2(g)$$

Therefore, CO_2 and H_2O are the products that should be written in the equation. The original equation then becomes

$$Na_2CO_3(aq) + 2\,HCl(aq) \longrightarrow 2\,NaCl(aq) + H_2O(l) + CO_2(g)$$

The evolution of a gas is indicated by a (g).

Examples of some other substances that decompose to form gases are sulfurous acid (H_2SO_3) and ammonium hydroxide (NH_4OH):

$$H_2SO_3(aq) \longrightarrow H_2O(l) + SO_2(g)$$
$$NH_4OH(aq) \longrightarrow H_2O(l) + NH_3(g)$$

Example 4. When solutions of sodium hydroxide and hydrochloric acid are mixed, the equation for the double displacement reaction (hypothetical) is

$$NaOH(aq) + HCl(aq) \longrightarrow NaCl(aq) + H_2O(l)$$

The mixture of these solutions produces no visible evidence of reaction, but on touching the test tube we notice that it feels warm. The evolution of heat is evidence of a chemical reaction. **Example 4** and **Example 1** appear similar because there is no visible evidence of reaction. However, the difference is very important. In **Example 1** all four ions are still uncombined. In **Example 4** the hydrogen ions (H^+) and hydroxide ions (OH^-) are no longer free in solution but have combined to form water. The reaction of H^+ (an acid) and OH^- (a base) is called **neutralization.** The formation of the slightly ionized compound (water) caused the reaction to occur and was the source of the heat liberated.

Water is the most common slightly ionized substance formed in double displacement reactions; other examples are acetic acid ($HC_2H_3O_2$), oxalic acid ($H_2C_2O_4$), and phosphoric acid (H_3PO_4).

From the four examples cited we see that a double displacement reaction will occur if at least one of the following classes of substances is formed by the reaction:

1. A precipitate

2. A gas

3. A slightly ionized compound, usually water

PROCEDURE

Wear protective glasses.

Each part of the experiment (except No. 12) consists of mixing equal volumes of two solutions in a test tube. Use about a **3 mL sample** of each solution (about 1.5 cm of liquid in a standard test tube). It is not necessary to measure each volume accurately. Record your observations at the time of mixing. Where there is no visible evidence of reaction, feel each tube, or check with a thermometer, to determine if heat is evolved (exothermic reaction). In each case where a reaction has occurred, complete and balance the equation, properly indicating precipitates and gases. When there is no evidence of reaction, write the words "No reaction" as the right-hand side of the equation.

1. Mix 0.1 M sodium chloride and 0.1 M potassium nitrate solutions.

2. Mix 0.1 M sodium chloride and 0.1 M silver nitrate solutions.

3. Mix 0.1 M sodium carbonate and **dilute.** (6 M) hydrochloric acid solutions.

4. Mix 10 percent sodium hydroxide and dil. (6 M) hydrochloric acid solutions.

5. Mix 0.1 M barium chloride and dil. (3 M) sulfuric acid solutions.

 6. Mix **dilute** (6 M) ammonium hydroxide and **dilute** (3 M) sulfuric acid solutions.

7. Mix 0.1 M copper(II) sulfate and 0.1 M zinc nitrate solutions.

8. Mix 0.1 M sodium carbonate and 0.1 M calcium chloride solutions.

9. Mix 0.1 M copper(II) sulfate and 0.1 M ammonium chloride solutions.

10. Mix 10 percent sodium hydroxide and dil. (6 M) nitric acid solutions.

11. Mix 0.1 M iron(III) chloride and dil. (6 M) ammonium hydroxide solutions.

 12. **Do this part in the fume hood.** Add 1 g of solid sodium sulfite to 3 mL of water and shake to dissolve. Now add about 1 mL of conc. (12 M) hydrochloric acid solution, a drop at a time, using a medicine dropper. Observe the results carefully.

WASTE DISPOSE OF PROPERLY Dispose of mixtures from reactions 2, 5, 7, 9 in the "heavy metal waste" container. Dispose of the contents of reaction 12 in the sink inside the hood. Dispose of the contents of all other tubes in the sink and flush with water.

REPORT FOR EXPERIMENT 11

Double Displacement Reactions

Directions for completing table below:

1. Record your observations (Evidence of Reaction) of each experiment. Use the following terminology: (a) "Precipitate formed" (include the color), (b) "Gas evolved," (c) "Heat evolved," or (d) "No reaction observed."

2. Complete and balance the equation for each case in which a reaction occurred. First write the correct formulas for the products, taking into account the charges (oxidation numbers) of the ions involved. Then balance the equation by placing a whole number in front of each formula (as needed) to adjust the number of atoms of each element so that they are the same on both sides of the equation. Use (g) or (s) to indicate gases and precipitates. Where no evidence of reaction was observed, write the words "No reaction" as the right-hand side of the equation.

Evidence of Reactions	Balanced Equations
1.	$NaCl$ + KNO_3 $\longrightarrow$
2.	$NaCl$ + $AgNO_3$ $\longrightarrow$
3.	Na_2CO_3 + HCl $\longrightarrow$
4.	$NaOH$ + HCl $\longrightarrow$
5.	$BaCl_2$ + H_2SO_4 $\longrightarrow$
6.	NH_4OH + H_2SO_4 $\longrightarrow$
7.	$CuSO_4$ + $Zn(NO_3)_2$ $\longrightarrow$
8.	Na_2CO_3 + $CaCl_2$ $\longrightarrow$
9.	$CuSO_4$ + NH_4Cl $\longrightarrow$
10.	$NaOH$ + HNO_3 $\longrightarrow$
11.	$FeCl_3$ + NH_4OH $\longrightarrow$
12.	Na_2SO_3 + HCl $\longrightarrow$

QUESTIONS AND PROBLEMS

1. The formation of what three classes of substances caused double displacement reactions to occur in this experiment?

   (a)

   (b)

   (c)

2. Write the equation for the decomposition of sulfurous acid.

3. Using three criteria for double displacement reactions, together with the Solubility Table in Appendix 5, predict whether a double displacement reaction will occur in each example below. If reaction will occur, complete and balance the equation, properly indicating gases and precipitates. If you believe no reaction will occur, write "no reaction" as the right-hand side of the equation. All reactants are in aqueous solution.

   (a) $K_2S + CuSO_4 \longrightarrow$

   (b) $NH_4OH + H_2C_2O_4 \longrightarrow$

   (c) $KOH + NH_4Cl \xrightarrow{\Delta}$

   (d) $NaC_2H_3O_2 + HCl \longrightarrow$

   (e) $Na_2CrO_4 + Pb(C_2H_3O_2)_2 \longrightarrow$

   (f) $(NH_4)_2SO_4 + NaCl \longrightarrow$

   (g) $BiCl_3 + NaOH \longrightarrow$

   (h) $KC_2H_3O_2 + CoSO_4 \longrightarrow$

   (i) $Na_2CO_3 + HNO_3 \longrightarrow$

   (j) $ZnBr_2 + K_3PO_4 \longrightarrow$

EXPERIMENT 12

Single Displacement Reactions

MATERIALS AND EQUIPMENT

Solids: strips of sheet copper, lead, and zinc measuring about 1×2 cm; and sandpaper or emery cloth. **Solutions:** 0.1 M copper(II) nitrate [$Cu(NO_3)_2$], 0.1 M lead(II) nitrate [$Pb(NO_3)_2$], 0.1 M magnesium sulfate ($MgSO_4$), 0.1 M silver nitrate ($AgNO_3$), and dilute (3 M) sulfuric acid (H_2SO_4). Small test tubes.

DISCUSSION

The chemical reactivity of elements varies over an immense range. Some, like sodium and fluorine, are so reactive that they are never found in the free or uncombined state in nature. Others, like xenon and platinum, are nearly inert and can be made to react with other elements only under special conditions.

The **reactivity** of an element is related to its tendency to lose or gain electrons; that is, to be oxidized or reduced. In principle it is possible to arrange nearly all the elements into a single series in order of their reactivities. A series of this kind indicates which free elements are capable of displacing other elements from their compounds. Such a list is known as an **activity** or **electromotive series.** To illustrate the preparation of an activity series, we will experiment with a small group of selected elements and their compounds.

A generalized single displacement reaction is represented by the equation

$$A(s) + BC(aq) \longrightarrow B(s) + AC(aq)$$

Element A is the more active element and replaces element B from the compound BC. But if element B is more active than element A, no reaction will occur.

Let us consider two specific examples, using copper and mercury.

Example 1. A few drops of mercury metal are added to a solution of copper(II) chloride ($CuCl_2$).

Example 2. A strip of metallic copper is immersed in a solution of mercury(II) chloride ($HgCl_2$).

In Example 1 no change is observed even after the solution has been standing for a prolonged time, and we conclude that there is no reaction. In Example 2 the copper strip is soon coated with metallic mercury, and the solution becomes pale green. From this evidence we conclude that mercury will not displace copper in copper compounds but copper will displace mercury in mercury compounds. Therefore copper is a more reactive metal than mercury and is above mercury in the activity series. In terms of chemical equations these facts may be represented as

Example 1. $Hg(l) + CuCl_2(aq) \longrightarrow$ No reaction

Example 2. $Cu(s) + HgCl_2(aq) \longrightarrow Hg(l) + CuCl_2(aq)$

The second equation shows that, in terms of oxidation numbers (or charges), the chloride ion remained unchanged, mercury changed from +2 to 0, and copper changed from 0 to +2. The +2 oxidation state of copper is the one normally formed in solution.

Expressed another way, the actual reaction that occurred was the displacement of a mercury ion by a copper atom. This can be expressed more simply in equation form:

$$Cu^0(s) + Hg^{2+}(aq) \longrightarrow Hg^0 + Cu^{2+}(aq)$$

In contrast to double displacement reactions, single displacement reactions involve changes in oxidation numbers and therefore are also classified as **oxidation-reduction reactions.**

PROCEDURE

Wear protective glasses.

1. Place six clean small test tubes in a rack and number them 1–6. To each, add about 2 mL of the solutions listed below.

2. Obtain three pieces of sheet zinc, two of copper, and one of lead. Be sure metal strips are small enough to fit into the test tubes. Clean the metal pieces with fine sandpaper or emery cloth to expose fresh metal surfaces. Add the metals to the test tubes with the solutions as listed.

   Tube 1: silver nitrate + copper strip
   Tube 2: copper(II) nitrate + lead strip
   Tube 3: lead(II) nitrate + zinc strip
   Tube 4: magnesium sulfate + zinc strip
   Tube 5: dilute (3M) sulfuric acid + copper strip
   Tube 6: dilute (3M) sulfuric acid + zinc strip

3. Observe the contents of each tube carefully and record any evidence of chemical reaction.

> Evidence of reaction will be either evolution of a gas (bubbles) or appearance of a metallic deposit on the surface of the metal strip. Metals deposited from a solution are often black or gray (in the case of copper, very dark reddish brown) and bear little resemblance to commercially prepared metals.
>
> With some of the combinations used in these experiments, the reactions may be slow or difficult to detect. If you see no immediate evidence of reaction, set the tube aside and allow it to stand for about 10 minutes, then reexamine it.

4. Pour the solutions in each test tube into the "heavy metals waste" container. Rinse the metals in tap water and dispose of the strips in the trash. Do not allow the metal strips to go into the sink or into the waste bottle.

NAME ____________________

SECTION __________ DATE __________

REPORT FOR EXPERIMENT 12

INSTRUCTOR ____________________

Single Displacement Reactions

Evidence of Reaction	**Equation (to be completed)**
Describe any evidence of reaction; if no reaction was observed, write "None".	Write "No reaction", if no reaction was observed.
1.	Cu + $AgNO_3(aq) \longrightarrow$
2.	Pb + $Cu(NO_3)_2(aq) \longrightarrow$
3.	Zn + $Pb(NO_3)_2(aq) \longrightarrow$
4.	Zn + $MgSO_4(aq) \longrightarrow$
5.	Cu + $H_2SO_4(aq) \longrightarrow$
6.	Zn + $H_2SO_4(aq) \longrightarrow$

QUESTIONS AND PROBLEMS

1. Complete the following table by writing the symbols of the two elements whose reactivities are being compared in each test:

	Tube Number					
	1	**2**	**3**	**4**	**5**	**6**
Greater activity						
Lesser activity						

2. Arrange Pb, Mg, and Zn in order of their activities, listing the most active first.

(1) __________

(2) __________

(3) __________

3. Arrange Cu, Ag, and Zn in order of their activities, listing the most active first.

(1) _____

(2) _____

(3) _____

4. Arrange Mg, H, and Ag in order of their activities, listing the most active first.

(1) _____

(2) _____

(3) _____

5. Arrange all five of the metals (excluding hydrogen) in an activity series, listing the most active first.

(1) _____

(2) _____

(3) _____

(4) _____

(5) _____

6. On the basis of the reactions observed in the six test tubes, explain why the position of hydrogen cannot be fixed exactly with respect to all of the other elements listed in the activity series in Question 5.

7. What additional test(s) would be needed to establish the exact position of hydrogen in the activity series of the elements listed in Question 5?

8. On the basis of the evidence developed in this experiment:

(a) Would silver react with dilute sulfuric acid? Why or why not?

(b) Would magnesium react with dilute sulfuric acid? Why or why not?

EXPERIMENT 13

Ionization-Electrolytes and pH

MATERIALS AND EQUIPMENT

Demonstration. **Solids:** sodium chloride (NaCl) and sugar ($C_{12}H_{22}O_{11}$). **Liquid:** glacial acetic acid ($HC_2H_3O_2$). **Solutions:** 0.1 M ammonium chloride (NH_4Cl), 1 M ammonium hydroxide (NH_4OH), 1 M acetic acid ($HC_2H_3O_2$), saturated barium hydroxide [$Ba(OH)_2$], 0.1 M copper(II) sulfate ($CuSO_4$), 1 M hydrochloric acid (HCl), 0.1 M nickel(II) nitrate [$Ni(NO_3)_2$], 0.1 M sodium bromide (NaBr), 1 M sodium hydroxide (NaOH), 0.1 M sodium nitrate ($NaNO_3$), and dilute (3 M) sulfuric acid (H_2SO_4). Conductivity apparatus; magnetic stirrer and stirring bar.

Solids: calcium hydroxide [$Ca(OH)_2$], calcium oxide (CaO), iron wire (paper clips), magnesium ribbon (Mg), magnesium oxide (MgO), marble chips ($CaCO_3$), sodium bicarbonate ($NaHCO_3$), sulfur (S), and wood splints. **Solutions:** dilute (6 M) acetic acid ($HC_2H_3O_2$), dilute (6 M) ammonium hydroxide (NH_4OH), dilute (6 M) hydrochloric acid (HCl), dilute (6 M) nitric acid (HNO_3), phenolphthalein, 10 percent sodium hydroxide (NaOH), and dilute (3 M) sulfuric acid (H_2SO_4). 0.001 M HCl, 0.01 M HCl, 0.1 M HCl for pH measurements, pH meter.

DISCUSSION

A. Electrolytes

Pure water will not conduct an electric current. However, when many solutes are dissolved in water, the resulting aqueous solutions will conduct electricity. These solutes, called **electrolytes,** form ions which are free to move in the solution. The electrical current through the solution is the movement of these ions to the positive and negative electrodes. Electrolytes are **acids, bases, and salts,** depending on the ions in solution. Other substances such as sugar and alcohol dissolve in water but are nonconductors because they do not form ions and are called **nonelectrolytes**.

The ions in an aqueous electrolyte solution are the result of the **dissociation** or **ionization** of compounds in water. Compounds that dissociate or ionize in water are **acids, bases and salts.** For example:

Dissociation of NaOH (a base) and NaCl (a salt):

$$NaOH(s) \xrightarrow{H_2O} Na^+(aq) + OH^-(aq)$$

$$NaCl(s) \xrightarrow{H_2O} Na^+(aq) + Cl^-(aq)$$

Ionization of HCl (a strong acid) and $HC_2H_3O_2$ (a weak acid)

$$HCl(g) + H_2O(l) \longrightarrow H_3O^+(aq) + Cl^-(aq)$$

$$HC_2H_3O_2(l) + H_2O(l) \rightleftharpoons H_3O^+(aq) + C_2H_3O_2^-(aq)$$

The necessity for water in this ionization process is illustrated by the fact that, when hydrogen chloride is dissolved in benzene, no ions are formed and the solution is a nonconductor (nonelectrolyte).

Electrolytes are classifed as strong or weak depending on the extent to which they exist as ions in solutions. **Strong electrolytes** are essentially 100 percent ionized in water, that is they exist totally as ions in solution. **Weak electrolytes** are considerably less ionized, only a small amount of the dissolved substance exists as ions, the remainder being in the un-ionized or molecular from. Most salts are strong electrolytes; acids and bases occur as both strong and weak electrolytes. Examples are as follows:

Strong Electrolytes	Weak Electrolytes
Most salts	$HC_2H_3O_2$
HCl	H_2SO_3
H_2SO_4	HNO_2
HNO_3	H_2CO_3
NaOH	H_2S
KOH	$H_2C_2O_4$
$Ba(OH)_2$	H_3PO_4
$Ca(OH)_2$	NH_4OH

In the first part of this experiment, the conductivity of many aqueous solutions will be demonstrated.

B. Acids

1. **Acids** are described as substances that yield hydrogen ions (H^+) when dissolved in water. This definition was first proposed by the Swedish chemist Arrhenius (over 100 years ago) for electrolytes which share common properties such as sour taste and the ability to change the color of the plant dye, litmus to red. This definition is the simplest way to think of acids and still applies.

Many compounds can be recognized as acids from their written formulas. The ionizable hydrogen atoms, which are responsible for the acidity, are written first, followed by the symbols of the other elements in the formula. Examples are:

HCl	Hydrochloric acid	H_2CO_3	Carbonic acid
HNO_3	Nitric Acid	HNO_2	Nitrous acid
H_2SO_4	Sulfuric Acid	H_2SO_3	Sulfurous Acid
$HC_2H_3O_2$	Acetic Acid	$H_2C_2O_4$	Oxalic Acid
H_3PO_4	Phosphoric Acid		

Acids are formed by the reaction of nonmetallic oxides called **acid anhydrides** with water. For example:

$$SO_3(g) + H_2O(l) \longrightarrow H_2SO_4(aq)$$

The chemical properties of acids will be observed in Procedure B.

2. **Bronsted-Lowry Acids and Bases**

The more inclusive Bronsted-Lowry acid-base theory defines acids as proton (H^+) donors and bases as proton acceptors. Thus, water behaves as both an acid and a base, as illustrated by the equation:

$$H_2O + H_2O \rightleftharpoons H_3O^+ + OH^-$$

$$\text{acid} \qquad \text{base} \qquad\qquad \text{acid} \qquad \text{base}$$

One water molecule has donated a proton, H^+, (acted as an acid) and another water molecule has accepted a proton (acted as a base). The hydronium ion, H_3O^+, is a hydrated hydrogen ion (H^+H_2O). To simplify writing equations, the formula of the hydronium ion is often abbreviated H^+. However, free hydrogen ions do not actually exist in aqueous solutions.

C. Bases

The Arrhenius definition for **bases** describes them as substances that yield hydroxide ions $(OH-)$ in water solutions. Bases change the color of litmus to blue. Common bases can be recognized by their formulas as a hydroxide ion (OH^-) combined with a metal or other positive ion. Examples are:

NaOH	Sodium hydroxide	KOH	Potassium hydroxide
$Ca(OH)_2$	Calcium hydroxide	$Mg(OH)_2$	Magnesium hydroxide
NH_4OH	Ammonium hydroxide		

The terms **alkali** and **alkaline** solutions are used synonymously with base and basic solutions.

Metal oxides that react with water to form bases are **basic anhydrides.** For example:

$$CaO(s) + H_2O(l) \longrightarrow Ca(OH)_2(aq)$$

The physical and chemical properties of bases will be observed in Procedure C.

D. Salts

Salts consist of a positively charged ion (H^+ excluded) and a negatively charged ion (O^{2-} and OH^- excluded). Salts may be formed by the reaction of acids and bases, or by replacing the hydrogen atoms in an acid with a metal, or by interaction of two other salts. There are many more salts than acids and bases. For example, for a single acid such as HCl we can produce many chloride salts (e.g. $NaCl$, KCl, $RbCl$, $CaCl_2$, NH_4Cl, $FeCl_3$, etc.)

The reaction of an acid and a base to form water and a salt is known as **neutralization.** For example:

$$HCl(aq) + NaOH(aq) \longrightarrow H_2O(l) + NaCl(aq)$$

E. The Importance and Measurement of H^+ Ion Concentration

An aqueous solution will be acidic, basic, or neutral, depending on the relative concentrations of H^+ and OH^-. In acidic solutions the concentration of the H^+ ions is greater than that of the OH^- ions. In basic solutions the concentration of the OH^- ions is greater than that of the H^+ ions. If the concentrations of H^+ and OH^- are equal (as in water), the solution is **neutral.**

There are two general methods for determining the relative concentrations of H^+ and OH^- and thus whether a solution is acid, alkaline, or neutral.

1. **Indicators** are organic compounds that change color at a particular hydrogen or hydroxide ion concentration. For example, litmus, a vegetable dye, shows a pink color in acidic solutions and a blue color in alkaline solutions. Another common indicator is phenolphthalein; it is colorless in acid solutions and pink in basic solutions. An indicator can only determine the relative concentrations of H^+ and OH^- within the range of its color changes.

2. A **pH meter** is an instrument designed so that it measures the H^+ directly and is used when an accurate measurement of the concentration of H^+ is needed. The pH meter is described and explained more fully in section (F).

F.   Measuring pH

The H^+ ion has a great effect on many chemical reactions, including biological processes that sustain life. For example, the H^+ concentration of human blood is regulated to very close tolerances. The concentration of this important ion is expressed as pH rather than other concentration expressions such as molarity. The pH is defined by this formula:

$$pH = -\log[H^+]$$

The H^+ written in brackets $[H^+]$ represents the concentration of H^+ in moles/liter. The logarithm (log) of a number is simply the power to which 10 must be raised to give that number. Thus, the log of 0.001 is -3 ($0.001 = 10^{-3}$). Since pH is defined as the negative log of an $[H^+]$ value, then the pH of a solution with $[H^+] = 0.001$ moles/liter is $-(-3)$ or pH = 3.

The pH of pure water is 7.0 at 25°C and is said to be neutral, that is, it is neither acidic nor basic because $[H^+]$ and $[OH^-]$ are equal (10^{-7} moles/liter). Solutions that are acidic have pH values less than 7.0. Solution that are basic have pH values greater than 7.0.

pH < 7.0	acid solutions
pH = 7.0	neutral solutions
pH > 7.0	basic (alkaline) solutions

A pH meter is a delicate instrument that comes in many versions and sizes which share some of these common features.

1. pH meters use a glass electrode that is immersed in the solution being tested. The electrode converts the H^+ concentration into an electrical potential which is read by a voltmeter calibrated in pH units.

2. pH meters are calibrated against standard solutions of known pH. After calibration, the electrode is immersed in the test solutions and the pH meter provides a value relative to the standard.

3. The operation of pH meters varies with different models. Your instructor will demonstrate the use of the pH meter available to you.

PROCEDURE

Wear protective glasses.

A. Conductivity of Solutions—Instructor Demonstration

All of the following tests (except number 8) are performed in 18×150 mm test tubes, using the conductivity apparatus shown in Figure 13.1 or other suitable conductivity apparatus. The electrodes should be rinsed thoroughly with distilled water between the testing of different solutions.

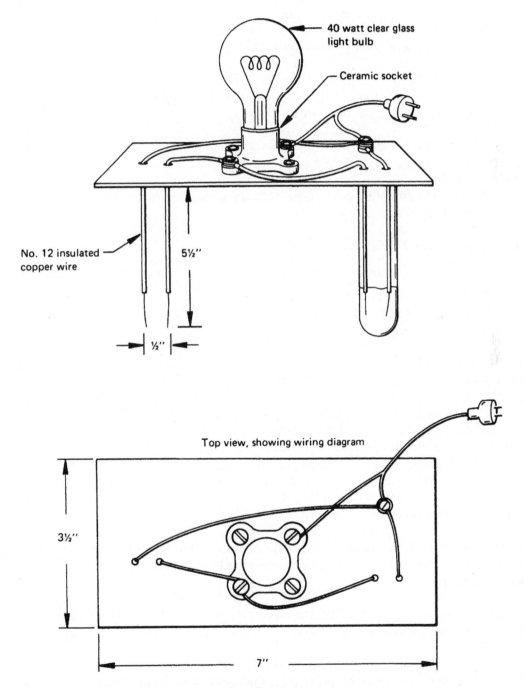

40 watt clear glass light bulb

Ceramic socket

No. 12 insulated copper wire

$5\frac{1}{2}''$

$\frac{1}{2}''$

Top view, showing wiring diagram

$3\frac{1}{2}''$

$7''$

Figure 13.1 Conductivity apparatus

Each test is performed by filling a test tube about half full of the liquid to be tested, then raising the test tube up around a pair of electrodes. When a measurable number of ions are in solution, the solution will conduct the electric current and the light will glow. A dimly glowing light indicates a relatively small number of ions in solution; a brightly glowing light indicates a relatively large number of ions in solution.

NOTE: The student should complete the data table in the report form at the time the demonstration is performed.

---

1. Test the conductivity of distilled water.

2. Test the conductivity of tap water.

3. Add a small amount of sugar to a test tube that is half full of distilled water. Dissolve the sugar and test the solution for conductivity.

4. Add a small amount of sodium chloride to a test tube that is half full of distilled water. Dissolve the salt and test the solution for conductivity.

 5. Remove the plug from the electrical outlet, clean and dry the electrodes, and reconnect the plug.

   (a) Test the conductivity of glacial acetic acid.

   (b) Pour out half of the acid, replace with distilled water, mix, and test the solution for conductivity.

   (c) Pour out half of the solution in 5(b), replace with distilled water, mix, and test the solution for conductivity.

6. Strong and weak acids and bases. Test the following 1 molar solutions for conductivity: (a) acetic acid, (b) hydrochloric acid, (c) ammonium hydroxide, (d) sodium hydroxide. If the conductivity apparatus has two sets of electrodes, as shown in Figure 13.1, the relative conductivity of the strong and weak acids or bases may be compared by alternately raising a tube of each solution around the electrodes. Clean and dry the electrodes (See No. 5).

7. Test the following 0.1 M salt solutions for conductivity: (a) sodium nitrate, (b) sodium bromide, (c) nickel(II) nitrate, (d) copper(II) sulfate, and (e) ammonium chloride.

8. Clean the electrodes well. Place about 25 mL of distilled water and 1 drop of dil. (3 M) sulfuric acid in a 150 mL beaker. Place the beaker on a magnetic stirrer and dip one set of electrodes into the solution. With the stirrer slowly turning, add saturated barium hydroxide solution dropwise until the light goes out completely. Add a few more milliliters of barium hydroxide solution.

B. Properties of Acids

 Dispose of all solutions in the sink and flush with water. Take care to make sure that solids such as metal strips, splints, and unreacted marble chips do not go into the sink. They should be put into the wastebasket.

1. **Reaction with a Metal**

   (a) Into four consecutive test tubes place about 5 mL of dil. (6 M) hydrochloric, (3 M) sulfuric, (6 M) nitric, and (6 M) acetic acids.

   (b) Place a small strip of magnesium ribbon into each tube, one at a time, and test the gas evolved for hydrogen by bringing a burning splint to the mouth of the tube. If the liberation of gas is slow, stopper the test tube loosely for a minute or two before testing for hydrogen.

2. **Measurement of Acidity and pH**

   (a) Test dilute solutions of hydrochloric acid, acetic acid, and sulfuric acid by placing a drop of each acid from a stirring rod onto a strip of red and onto a strip of blue litmus paper. Note any color changes.

   (b) Add 2 drops of phenolphthalein solution to about 5 mL of distilled water. Add several drops of dilute hydrochloric acid, mix, and note any color change.

   (c) Use the pH meter to measure the pH of three dilutions of hydrochloric acid in this order: 0.001 M HCl, 0.01 M HCl, and 0.1 M HCl. *Rinse the electrodes thoroughly with distilled water when done.*

3. **Reaction with Carbonates and Bicarbonates**

   (a) Cover the bottom of a 150 mL beaker with a small quantity of sodium bicarbonate powder. Now add about 4 to 5 mL of dil. (6 M) hydrochloric acid to the beaker and cover with a glass plate. After about 30 seconds lower a burning splint into the beaker and observe the results. Dispose of the reaction mixture in the sink.

   (b) Repeat the above experiment, using a few granules of marble chips (calcium carbonate) instead of sodium bicarbonate. Allow the reaction to proceed for 2 minutes before testing with the burning splint. Dispose of unreacted marble chips in the wastebasket, not the sink.

4. **Reaction with Bases—Neutralization.** To about 25 mL of water in a beaker, add 3 drops of phenolphthalein solution and 5 drops of dil. (6 M) hydrochloric acid. Using a medicine dropper, add 10 percent sodium hydroxide solution dropwise, stirring after each drop, until the indicator in the solution changes color. Then add dilute hydrochloric acid, drop by drop, stirring after each drop, until the indicator becomes colorless again. Repeat the additions of base and acid one or two more times. Dispose of all solutions in the sink.

5. **Nonmetal Oxide plus Water** Dispose of all solutions in the sink.

   (a) **Do this part in the fume hood.** Place a small lump of sulfur in a deflagrating spoon and start it burning by heating in the burner flame. Lower the burning sulfur into a wide-mouth bottle containing 15 mL of distilled water and let the sulfur burn for 2 minutes. Remove the deflagrating spoon and quench the excess burning sulfur in a beaker of water. Cover the bottle with a glass plate and shake the bottle back and forth to dissolve the sulfur dioxide gas. Test the solution with blue litmus paper.

   (b) As shown in Figure 13.2, fit a test tube with a one-hole stopper containing a glass delivery tube long enough to extend to the bottom of another test tube. Place several pieces of marble chips and a few milliliters of dil. (6 M) hydrochloric acid into the tube and insert the stopper. Bubble the liberated carbon dioxide into another test tube containing 10 mL water, 1 drop of 10 percent sodium hydroxide solution, and 2 drops of phenolphthalein solution. Record the results.

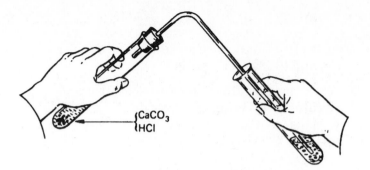

Figure 13.2   Generator for carbon dioxide

C.   Properties of Bases

Dispose of all solutions in the sink.

1. "Feel" Test. Make very dilute solutions by adding 5 drops of dilute (6 M) ammonium hydroxide to 10 mL of water in a test tube and 3 drops of 10 percent sodium hydroxide solution to 10 mL of water in another test tube. Rub a small amount of each very dilute solution between your fingers to obtain the characteristic "feel" of a hydroxide (base) solution. Wash your hands thoroughly immediately after making the "feel" test. Save the very dilute base solutions for the measurement of pH in the next section, C2(c).

2. **Measurement of Alkalinity**

   (a) Test the two base (alkaline) solutions prepared in C.1 with both red and blue litmus paper. Note any color changes.

   (b) Add 2 drops of phenolphthalein solution to each of the two alkaline solutions prepared in C.1. Note any color changes.

   (c) Pour the dilute ammonium hydroxide and sodium hydroxide that were prepared in the previous step into separate small beakers. Use the pH meter to measure the pH of these alkaline solutions. *Dip the electrode in dilute acetic acid and rinse thoroughly with distilled water when done.*

3. **Metal Oxides plus Water**

   (a) Place 10 mL of water and 2 drops of phenolphthalein solution in each of 3 test tubes. Add a pinch of calcium oxide to the first, magnesium oxide to the second, and calcium hydroxide to the third tube. Note and record the results.

   (b) Wind the end of a 5 cm piece of iron wire (or paper clip) around a small marble chip. Grasp the wire with tongs and heat the marble chip in the hottest part of the burner (flame for about 2 minutes—the edges of the chip should become white hot while being heated. Allow the chip to cool; then drop it into a beaker containing 15 mL of water and 2 drops of phenolphthalein solution. For comparison, repeat this part of the experiment with a marble chip which has not been heated. Note the results. **Return the iron wire to the reagent shelf.** Dispose marble chips in the wastebasket.

4. **Reaction with Acids—Neutralization.** Review Part B.4.

REPORT FOR EXPERIMENT 13

Ionization—Electrolytes and pH

A.   Conductivity of Solutions—Instructor Demonstration

Complete the table for each of the substances tested in the ionization demonstration. Place an "X" in the column where the property of the substance tested fits the column description.

	Nonelectrolyte	Strong Electrolyte	Weak Electrolyte
1. Distilled Water			
2. Tap water			
3. Sugar			
4. NaCl			
5. a. $HC_2H_3O_2$ (glacial)			
b. 1st dilution			
c. 2nd dilution			
6. a. 1 M $HC_2H_3O_2$			
b. 1 M HCl			
c. 1 M NH_4OH			
d. 1 M NaOH			
7. a. $NaNO_3$			
b. NaBr			
c. $Ni(NO_3)_2$			
d. $CuSO_4$			
e. NH_4Cl			

8. (a) Write an equation for the chemical reaction that occurred between sulfuric acid and barium hydroxide.

(b) Explain in terms of the properties of the products formed why the light went out when barium hydroxide was added to sulfuric acid solution, even though both of these reactants are electrolytes.

(c) Explain why the light came on again when additional barium hydroxide was added.

9. In the conductivity tests, what controlled the brightness of the light?

10. Write an equation to show how acetic acid reacts with water to produce ions in solution.

11. What classes of compounds tested are electrolytes?

B. Properties of Acids

1. Reaction with a Metal

(a) Write the formulas of the acids which liberated hydrogen gas when reacting with magnesium metal.

_____

(b) Write equations to represent the reactions in which hydrogen gas was formed.

2. **Measurement of Acidity and pH**

   (a) What is the effect of acids on the color of red litmus?

   (b) What is the effect of acids on the color of blue litmus?

   (c) What color is phenolphthalein in an acid solution?   _____

   (d) What was the pH of the hydrochloric acids tested?

       0.001 M _____   0.01 M _____   0.1 M _____

   (e) Which pH measured has the highest number of H^+ in solution?   _____

   (f) What is the H^+ concentration in an acid with of pH 4.6?
   Express your answer as a power of 10   _____

       Refer to Study Aid 4 if you need help with using your calculator to convert the pH into H^+ concentration using the antilog function.

3. **Reaction with Carbonates and Bicarbonates**

   (a) What gas is formed in these reactions?

       Name _____   Formula _____

   (b) What happened to the burning splint when it was thrust into the beaker?

   (c) What do you conclude about one of the properties of the gas in the beaker, based on the behavior of the burning splint?

   (d) Complete and balance the equations representing the reactions:

       $NaHCO_3(s) + HCl(aq) \longrightarrow$

       $CaCO_3(s) + HCl(aq) \longrightarrow$

   4. **Reaction of Acids with Bases—Neutralization**

      (a) Write an equation for the neutralization reaction of HCl and NaOH.

      (b) How did you know when all the acid was neutralized?

   5. **Nonmetal Oxide plus Water**

      (a) Write an equation for the combustion of sulfur in air.

      (b) What acid is formed when the product of the sulfur combustion reacts with water?

         Name _____   Formula _____

         Write the equation for its formation.

      (c) What evidence in this experiment leads you to believe that carbon dioxide in water has acidic properties?

      (d) What acid is formed when carbon dioxide reacts with water?

         Name _____   Formula _____

C.   **Properties of Bases**

  1. **"Feel" Test.** What is the characteristic feel of basic solutions?

  2. **Measurement of Alkalinity**

    (a) What is the effect of bases on the color of red litmus?

(b) What is the effect of bases on the color of blue litmus?

(c) What color is phenolphthalein in a basic solution? _____

(d) What was the pH for each dilute base tested?

$NH_4OH(aq)$ _____ $NaOH(aq)$ _____

(e) Which base tested has the highest number of H^+ in solution? _____

(f) What is the H^+ concentration in the strongest base tested?
Express your answer as a power of 10. _____

Refer to Study Aid 4 if you need help using your calculator to convert the pH into H^+ concentration using the antilog function.

3. **Metal Oxides plus Water**

(a.1) Color (if any) produced by phenolphthalein.

Color with CaO in water _____

Color with MgO in water _____

Color with $Ca(OH)_2$ in water _____

(a.2) Complete and balance these equations:

$CaO(s) + \quad H_2O(l) \longrightarrow$

$MgO(s) + \quad H_2O(l) \longrightarrow$

(b.1) The formula for marble is $CaCO_3$, What compounds are formed when it is heated strongly?

(b.2) Write the equation representing this decomposition:

$CaCO_3(s) \xrightarrow{\Delta}$

(b.3) What evidence led you to formulate the composition of the solid residue after heating the marble chip?

ADDITIONAL QUESTIONS AND PROBLEMS

1. State whether each of the formulas below represents an **acid,** a **base,** a **salt,** an **acid anhydride,** a **basic anhydride,** or **none** of these types of compounds:

CuF_2 _____ $CaSO_4$ _____

$Ba(OH)_2$ _____ C_2H_4 _____

$LiOH$ _____ $C_{12}H_{22}O_{11}$ _____

$HBrO_3$ _____ HI _____

$RaCO_3$ _____ P_2O_5 _____

KNO_2 _____ HCN _____

$H_2C_2O_4$ _____ MgO _____

2. Complete and balance the following equations and name the product formed. (Only one product is formed in each case.)

Name of Product

(a) $K_2O(s) + H_2O(l) \longrightarrow$ _____

(b) $SrO(s) + H_2O(l) \longrightarrow$ _____

(c) $SO_3(s) + H_2O(l) \longrightarrow$ _____

(d) $N_2O_5(s) + H_2O(l) \longrightarrow$ _____

EXPERIMENT 14

Identification of Selected Anions

MATERIALS AND EQUIPMENT

Liquids: Decane ($C_{10}H_{22}$). **Solutions:** 0.1 M barium chloride ($BaCl_2$), freshly prepared chlorine water (Cl_2), dilute (6 M) hydrochloric acid (HCl), dilute (6 M) nitric acid (HNO_3), 0.1 M silver nitrate ($AgNO_3$), 0.1 M sodium bromide (NaBr), 0.1 M sodium carbonate (Na_2CO_3), 0.1 M sodium chloride (NaCl), 0.1 M sodium iodide (NaI), 0.1 M sodium phosphate (Na_3PO_4), 0.1 M sodium sulfate (Na_2SO_4), and unknown solutions. Wash bottle for distilled water.

DISCUSSION

The examination of a sample of inorganic material to identify the ions that are present is called **qualitative analysis.** To introduce qualitative analysis, we will analyze for six anions (negatively charged ions). The ions selected for identification are chloride (Cl^-), bromide (Br^-), iodide (I^-), sulfate ($SO_4{}^{2-}$), phosphate ($PO_4{}^{3-}$) and carbonate ($CO_3{}^{2-}$).

Qualitative analysis is based on the fact that no two ions behave identically in all of their chemical reactions. Identification depends on appropriate chemical tests coupled with careful observation of such characteristics as solution color, formation and color of precipitates, evolution of gases, etc. Test reactions are selected to identify the ions in the fewest steps possible. In this experiment only one anion is assumed to be present in each sample. If two or more anions must be detected in a single solution, the scheme of analysis can be considerably more complex.

Silver Nitrate Test

When solutions of the sodium salts of the six anions are reacted with silver nitrate solution, the following precipitates are formed: AgCl, AgBr, AgI, Ag_3PO_4, and Ag_2CO_3. Ag_2SO_4 is moderately soluble and does not precipitate at the concentrations used in these solutions. When dilute nitric acid is added, the precipitates Ag_3PO_4, and Ag_2CO_3 dissolve; AgCl, AgBr, and AgI remain undissolved. Acids react with carbonates to form CO_2 (g). Look for gas bubbles when nitric acid is added to the silver precipitates.

In some cases a tentative identification of an anion may be made from the silver nitrate test. This identification is based on the color of the precipitate and on whether or not the precipitate is soluble in nitric acid. However, since two or more anions may give similar results, second or third confirmatory tests are necessary for positive identification.

Barium Chloride Test

When barium chloride solution is added to solutions of the sodium salts of the six anions, precipitates of $BaSO_4$, $Ba_3(PO_4)_2$ and $BaCO_3$, are obtained. No precipitate is obtained with Cl^-, Br^-, or I^-.

When dilute hydrochloric acid is added, the precipitates $Ba_3(PO_4)_2$ and $BaCO_3$ dissolve; $BaSO_4$ does not dissolve. Look for CO_2 gas bubbles.

Organic Solvent Test

The silver nitrate test can prove the presence of a halide ion (Cl^-, Br^-, or I^-) because the silver precipitates of the other three anions dissolve in nitric acid. But the colors of the three silver halides do not differ sufficiently to establish which halide ion is present.

Adding chlorine water (Cl_2 dissolved in water) to halide salts in solution will oxidize bromide ion to free bromine (Br_2) and iodine ion to free iodine (I_2). The free halogen may be extracted from the water solution by adding an immiscible organic solvent such as decane and shaking vigorously. The colors of the three halogens in organic solvents are quite different. Cl_2 is pale yellow, Br_2 is yellow-orange to reddish-brown, and I_2 is pink to violet. After adding chlorine water and shaking, a yellow-orange to reddish-brown color in the decane layer indicates that Br^- was present in the original solution; a pink to violet color in the decane layer indicates that I^- was present. However, a pale yellow color does not indicate Cl^-, since Cl_2 was added as a reagent. But if the silver nitrate test gives a white precipitate that is insoluble in nitric acid, and the organic solvent test shows no Br^- or I^-, then you can conclude that Cl^- was present.

Though we have described many of the expected results of these tests, it is necessary to test known solutions to actually see the results of the tests and to develop satisfactory experimental techniques. During this experiment, you will perform these tests on six known anions.

Then, two "unknown" solutions, each containing one of the six anions, will be analyzed. When an unknown is analyzed, the results should agree in all respects with one of the known anions. If the results do not fully agree with one of the six known ions, either the testing has been poorly done or the unknown does not contain any of the specified ions.

Three different kinds of equations may be used to express the behavior of ions in solution. For example, the reaction of the chloride ion (from sodium chloride) may be written.

1. $NaCl(aq) + AgNO_3(aq) \longrightarrow AgCl(s) + NaNO_3(aq)$

2. $Na^+(aq) + Cl^-(aq) + Ag^+(aq) + NO_3^-(aq) \longrightarrow AgCl(s) + Na^+(aq) + NO_3^-(aq)$

3. $Cl^-(aq) + Ag^+(aq) \longrightarrow AgCl(s)$

Equation (1) is the **formula (un-ionized) equation;** it shows the formulas of the substances in the equation as they are normally written. Equation (2) is the **total ionic equation;** it shows the substances as they occur in solution. Strong electrolytes are written as ions; weak electrolytes, precipitates, and gases are written in their un-ionized or molecular form. Equation (3) is the **net ionic equation;** it includes only those substances or ions in Equation (2) that have undergone a chemical change. Thus Na^+ and NO_3^- (sometimes called the "spectator" ions) have not changed and do not appear in the net ionic equation. In both the total ionic and net ionic equations, the atoms and charges must be balanced.

PROCEDURE

Wear protective glasses

1. Clean eight test tubes and rinse each twice with 5 mL of distilled water. The first six test tubes are for the known solutions that will be tested to demonstrate the expected reactions with each anion. Use a marker to label these tubes as follows: NaCl, NaBr, NaI, Na_2SO_4, Na_3PO_4 and Na_2CO_3. The last two tubes are for your unknowns and should be left blank for now. Arrange these test tubes in order in your test tube rack.

2. Clean and rinse two more test tubes and take them to your instructor for your unknown solutions and their identification code. Label them with the code numbers immediately. To avoid possible confusion with the empty unknown test tubes in the rack, put these coded tubes aside in a beaker. Record the code of these unknowns in the top right-hand columns of your report form and label each of the blank tubes in the rack with one of these unknown code numbers.

Pour 2 mL (no more) of each of the six known solutions—one solution per tube—and 2 mL of the corresponding unknown into each unknown tube. Save the remaining portions of the unknown solutions for tests B and C.

You can save considerable time by measuring out 2 mL into the first test tube and using the height of this liquid in the test tube as a guide for measuring out the others.

Dispose of solutions containing decane in the container marked "Waste organic solvents." Dispose of solutions containing silver, and barium, in the "heavy metals waste" container.

> For each of the following tests that will be performed on known and unknown solutions, there is a corresponding block on the report form where observations should be recorded. If a precipitate forms, record "ppt formed" and include its color. If no precipitate forms, record "no ppt." When dissolving precipitates, record "ppt dissolved" or "ppt did not dissolve." For the decane solubility test, indicate the color of the decane layer.

A. Silver Nitrate Test

Silver nitrate will stain your skin black. If any silver nitrate gets on your hands, wash it off immediately to avoid these stains.

Add about 1 mL of 0.1 M silver nitrate solution to each test tube. Record the results. Now add about 3 mL of dilute (6 M) nitric acid to each test tube; stopper and shake well. Record the results.

B. Barium Chloride Test

Wash all eight test tubes and rinse each tube twice with distilled water. Again put about 2 mL of the specified solution into each of the eight test tubes. Add about 2 mL of 0.1 M barium chloride solution to each test tube and mix. Record the results. Now add 3 mL of dilute hydrochloric acid to each tube; stopper and shake well. Record the results.

C. Organic Solvent Test

Again wash and rinse all eight test tubes. Again put about 2 mL of the specified solution into each of the eight test tubes. Now add about 2 mL of decane and about 2 mL of chlorine water to each test tube; stopper and shake well. Record the results.

After completing the three tests, compare the results of the known solutions with your observations for your unknown solutions. Record the formula of the anion present in each solution on the report form (Part D).

REPORT FOR EXPERIMENT 14

Identification of Selected Anions

	NaCl	NaBr	NaI	Na_2SO_4	Na_3PO_4	Na_2CO_3	Unknown No. ____	Unknown No. ____
A. AgNO$_3$ Test Addition of AgNO$_3$ solution								
Addition of dil. HNO$_3$								
B. BaCl$_2$ Test Addition of BaCl$_2$ solution								
Addition of dil. HCL								
C. Organic Solvent Test Color of decane layer								
D. Formula of anion present in the solution tested.								

QUESTIONS AND PROBLEMS

1. The following three solutions were analyzed according to the scheme used in this experiment. Which one, if any, of the ions tested, is present in each solution? If the data indicate that none of the six is present, write the word "None" as your answer.

   (a) **Silver Nitrate Test.**   Yellow precipitate formed, which dissolved in dilute nitric acid.

   Barium Chloride Test.   White precipitate formed, which dissolved in dilute hydrochloric acid.

   Organic Solvent Test.   The decane layer remained almost colorless after treatment with chlorine water.

   Anion present _____

   (b) **Silver Nitrate Test.**   Red precipitate formed, which dissolved in dilute nitric acid to give an orange solution.

   Barium Chloride Test.   Yellow precipitate formed, which dissolved in dilute hydrochloric acid to give an orange solution.

   Organic Solvent Test.   The decane layer remained almost colorless after treatment with chlorine water.

   Anion present _____

   (c) **Silver Nitrate Test.**   Yellow precipitate formed, which did not dissolve in dilute nitric acid.

   Barium Chloride Test.   No precipitate formed.

   Organic Solvent Test.   The decane layer turned reddish-brown.

   Anion present _____

2. Write formula, total ionic, and net ionic equations for the following reactions: Use the solubility table in Appendix 5 for reactions that were not observed directly in this experiment. All reactions are in aqueous solutions.

(a) Sodium bromide and silver nitrate.

(b) Sodium carbonate and silver nitrate.

(c) Sodium arsenate and barium chloride.

3. Write net ionic equations for the following reactions. Assume that a precipitate is formed in each case.

(a) Sodium iodide and silver nitrate.

(b) Sodium acetate and silver nitrate.

(c) Sodium phosphate and barium chloride.

(d) Sodium sulfate and barium chloride.

EXPERIMENT 15

Quantitative Preparation of Potassium Chloride

MATERIALS AND EQUIPMENT

Solid: potassium bicarbonate ($KHCO_3$). **Solution:** 6 M hydrochloric acid (HCl).

DISCUSSION

In this experiment you will examine and verify the mole and mass relationships involved in the quantitative preparation of potassium chloride. Potassium bicarbonate is the source of the potassium ion, and hydrochloric acid is the source of chloride ions. The reaction is expressed in the following equation, which shows that potassium bicarbonate and hydrochloric acid react with each other in a 1-to-1-mole ratio:

$$KHCO_3(aq) + HCl(aq) \longrightarrow KCl(aq) + H_2O(l) + CO_2(g)$$

Furthermore, for every mole of potassium bicarbonate present, 1 mole of potassium chloride is formed. From these molar relationships we can calculate the amount of potassium chloride that is theoretically obtainable from any specified amount of potassium bicarbonate in the reaction. The experimental value can then be compared to the theoretical value.

To conduct the experiment quantitatively, we need to react all the potassium ion from a known amount of potassium bicarbonate and to isolate the KCl in pure a form as feasible. To ensure complete reaction of the potassium bicarbonate, an excess of hydrochloric acid is used The end of the reaction is detectable because the evolution of the gaseous product CO_2 stops when all the $KHCO_3$ has been reacted.

Use the following relationships in your calculations:

1. 1 mole $KHCO_3$ reacted = 1 mole HCl reacted = 1 mole KCl produced

2. 1 mole of solute = 1 molar mass of solute

     Example: moles $KHCO_3$ = (g $KHCO_3$)$\left(\dfrac{1 \text{ mol } KHCO_3}{100.1 \text{ g } KHCO_3} \right)$

3. Molarity = $\dfrac{\text{moles solute}}{\text{L solution}}$ and for the HCl used in this reaction we can set up the conversion factors

$$\frac{6.0 \text{ mol HCl}}{1 \text{ L}} \quad \text{or} \quad \frac{6.0 \text{ mol HCl}}{1000 \text{ mL}}$$

Note that molarity is an expression of concentration, the units of which are *always* moles of solute per liter of solution from which conversions factors for mol $\longleftrightarrow$ volume can be derived.

For example, if you wanted to determine the volume of 2.0 M HCl that would be used to complete the reaction with 5.5000 g of KHCO$_3$, the dimensional analysis setup would be:

$$\text{mL HCl} = (5.5000 \,\text{g KHCO}_3)\left(\frac{1 \,\text{mole KHCO}_3}{100.1 \,\text{g KHCO}_3}\right)\left(\frac{1 \,\text{mol HCl}}{1 \,\text{mol KHCO}_3}\right)\left(\frac{1000 \,\text{mL}}{2.0 \,\text{mol HCl}}\right) = 27 \,\text{mL HCl}$$

4. Percentage error $= \left(\dfrac{\text{theoretical value} - \text{experimental value}}{\text{theoretical value}}\right)(100)$

The sequence of major experimental steps in this experiment is as follows:

1. Weigh an evaporating dish.

2. Weigh 2-3 g potassium bicarbonate into the evaporating dish.

3. Dissolve potassium bicarbonate in 5 mL distilled water.

4. Add hydrocholoric acid solution slowly until the fizzing stops.

5. Evaporate the liquid to obtain the dry product, KCl.

6. Heat and dry the KCl to constant weight.

7. Determine the mass of KCl produced.

PROCEDURE

Wear protective glasses.

---

1. Make all weighings *to the highest precision* possible with the balance available.

2. Use the same balance for all weighings.

3. Record all data directly on the report form as they are obtained.

---

1. Weigh a clean, dry evaporating dish.

2. Now add between 2 and 3 g (no more) of potassium bicarbonate to the evaporating dish and reweigh.

3. Dissolve the potassium bicarbonate in 5 mL of distilled water. If all the potassium bicarbonate does not completely dissolve, do not worry about it. Continue on with the next step.

4. In a graduated cylinder, obtain 6.0 mL of 6 M HCl and **slowly, *with* stirring,** add it to the bicarbonate solution. (The product is formed in this step).

5. Using a beaker of water to make a water bath as shown in Experiment 1, Part C, evaporate the liquid from the solution of potassium chloride. Replenish the water in the water bath as needed. When the water has essentially evaporated (the residue in the dish

looks dry), allow the system to cool for a few minutes; remove the evaporating dish and thoroughly dry the bottom of the dish.

6. The following method of drying the product must be followed to avoid spattering and loss of product. Pay attention during this procedure. Do not leave the drying setup unattended.

   Adjust the burner so you have a nonluminous, 10 to 15 cm (4 to 6 in.) flame **without a distinct inner cone.** Place the evaporating dish on a wire gauze 4-6 in. above the top of the barrel. Heat the dish and contents for 5-10 minutes (the KCl should appear dry). Touch the surface with a stirring rod to prevent the formation of a crust. If spattering occurs remove the burner momentarily and either lower the flame or raise the dish before continuing heating.

7. Cool the dish, weigh, and reheat for an additional 5 minutes. Cool again and reweigh. If the second weighing is within 0.08 g of the first, the KCl may be considered dry. If the second weighing has decreased more than 0.08 g, a third heating (5 minutes) and weighing is necessary. The experiment is complete after obtaining constant weight (within 0.08 g). If constant weight is not obtained after three heatings, your instructor will provide instructions on what to do.

8. From the data collected, determine the mass of KCl produced.

 9. Dissolve the KCl in water and wash it down the sink.

REPORT FOR EXPERIMENT 15

Quantitative Preparation of Potassium Chloride

A. Write the balanced equation for the reaction between $KHCO_3$ and HCl:

B. Experimental Data and Calculations: Record all measurement to the highest precision of the balance and remember to use the proper number of significant figures in all calculations. (The number 0.004 has only *one* significant figure.)

    1. Mass of empty evaporating dish   _____ g

    2. Mass of dish and dry $KHCO_3$ _____ g

    3. Mass of dish and residue (KCl) after first heating   _____ g

    4. Mass of dish and residue (KCl) after second heating   _____ g

    5. Mass of dish and residue (KCl) after third heating (if necessary)   _____ g

    6. Mass of potassium bicarbonate   _____ g
        show calculation set-up

    7. Moles of potassium bicarbonate   _____ mol
        show calculation set-up

    8. Experimental mass of potassium chloride obtained   _____ g
        show calculation set-up

    9. Experimental moles of potassium chloride obtained   _____ mol
        show calculation set-up

    10. Theoretical moles of KCl   _____ mol
        show calculation set-up

    11. Theoretical mass of KCl   _____ g
        show calculation set-up

    12. Percentage error for experimental mass of KCl vs.   _____ %
        theoretical mass of KCl (show calculation set-up)

QUESTIONS AND PROBLEMS

1. What was done in the experiment to make sure that all the $KHCO_3$ was reacted?

2. Why is the mass of KCl recovered less than the starting mass of $KHCO_3$?

3. Calculate the moles and grams of HCl present in the 6.0 mL of 6.0 M HCl solution you used.

                                                         _____ mol HCl

                                                         _____ g HCl

4. Would the 6.0 mL of 6.0 M HCl be sufficient to react with 3.80 g $KHCO_3$? Show supporting calculations and explanation.

5. Theoretically, why should the moles of $KHCO_3$ and the moles of KCl produced be the same?

6. If 3.000 g of K_2CO_3 were used in this experiment (instead of $KHCO_3$),

    (a) What is the balanced equation for the reaction?

    (b) How many milliliters of 6.0 M HCl would be needed   _____ mL HCl

    (c) How many grams of KCl would be formed in the reaction?   _____ g KCl

EXPERIMENT 16

Electromagnetic Energy and Spectroscopy

MATERIALS AND EQUIPMENT

Solutions: 0.1 M Nickel(II) nitrate, $Ni(NO_3)_2$; 0.002 M potassium permanganate, $KMnO_4$.
Special equipment: Hand-held spectroscopes; 1.75 m long springs for simulating wave motion (CENCO #84740G); meter sticks; spectrophotometer and cuvettes; vapor lamps (hydrogen, neon) with power supplies; incandescent 60-100 watt bulbs, fluorescent light, spectrum chart, colored pencils, stopwatch.

DISCUSSION

Background: Electron Arrangements and Electromagnetic Energy

All substances will emit or absorb electromagnetic (EM) energy in a unique manner. You are most familiar with a range of EM energy known as visible light. A common example of emitted visible light energy is the red glow of a neon sign. A common example of the effects of absorbed visible light is the green color of the plant pigment, chlorophyll. The patterns of emitted and absorbed light by substances are the basis on which Neils Bohr and others built the model of atomic structure which is accepted today. In this model, the neutrons (no charge) and protons (positive charge) are packed tightly into a dense nucleus. The electrons are arranged in energy levels with the lowest energy electrons closest to the nucleus and the most energetic electrons farthest away. This can be shown in a diagram similar to the one below for sulfur.

increasing energy level $\longrightarrow$

$$n= \quad 1 \quad\quad 2 \quad\quad 3$$

$\overbrace{\binom{16p}{17n}}$ $\quad 2e^- \quad 8e^- \quad 6e^-$

The hydrogen atom has only one proton in its nucleus (no neutrons) and one electron which will be in the first energy level when the atom is in its most stable configuration, called the ground state. When the atom absorbs energy this electron becomes excited and moves farther away from the nucleus into a higher energy level. Eventually, the excited electron loses energy and returns to a lower energy level, emitting energy during the transition. The absorption of certain wavelengths of emitted energy by the bonds of pigment molecules in the retina enable us to perceive light and color.

For the hydrogen electron, some of the possible transitions for an excited electron moving closer to its ground state are shown in Figure 16.1. If the transition of the electron is into energy level 2, the emitted energy fits the absorption spectrum of the vision pigments. For hydrogen, there are 4 such transitions. If the transition is into energy level 1 from a higher energy level, the emitted energy is not absorbed by the electrons of vision molecules and, although energetic and significant, the energy is not visible. If the transition is into energy level 3, the emitted energy is also not absorbed by vision pigments.

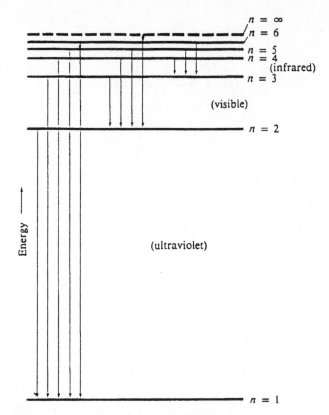

Figure 16.1   Electron transitions in the H atom

A.   Wave Properties of Electromagnetic Energy

Because only certain values of EM energy can be absorbed for any particular atom, the absorption is said to be quantized. Because of this, any change in the electronic energy level of an atom involves the absorption or emission of a definite amount or quantum of energy. These packets or quanta of energy are called photons. They are emitted or absorbed by electrons in an atom as they change energy levels; and as they travel through space, they are referred to as **electromagnetic radiation**. All photons of light, regardless of their energy content, exhibit wave-like behavior and travel at the same speed in a vacuum (3.00×10^8 m/s). Therefore, photons are often described by their wave properties, specifically, their **wavelength (lambda, λ)** and their frequency (nu, ν).

As photons move through space they do not move in a smooth straight line. Instead, their straight-line path is displaced slightly from its position (depending on its energy content) and once displaced, it tends to correct itself, returning to its original position, and then overcorrect itself. In this process, a wave pattern in generated. Light waves are described as transverse and have particle displacement perpendicular to the motion of the wave. The resultant wave is usually drawn as shown in Figure 16.2.

The wavelength of these waves is shown by λ and can be measured from peak to peak or trough to trough. Frequency tells how many waves pass a particular point per second. Depending on the energy content of the photon, the wave properties of a given particle of energy will vary greatly. Thus, there are photons with very short wavelengths (10^{-12} m) and very long wavelengths (10^4 m), or with very high frequencies (10^{20} waves/s) or with very low frequencies (10^9 waves/s). The relationship between wavelength, energy, and frequency will be demonstrated in the first activity of this experiment.

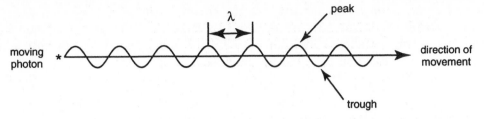

Figure 16.2 Transverse wave pattern of EM energy

B. Emission Spectra

Photons of energy are released or emitted by excited electrons as they fall back to lower energy levels. The range of photons characteristic of a given atom as its excited electrons fall closer to the nucleus is called its **emission spectrum.** Photons with a wavelength between 350 and 650 nm are absorbed by electrons in the retinal molecules of the eye which initiates a chain of events leading to their perception in the brain as visible light. Thus, these photons are called the visible light spectrum. A special filter or prism, called a **spectroscope,** can be used to separate photons of various wavelengths emitted by a given source so they can be identified. If

C. Absorption Spectra

Just as atoms emit characteristic photons of energy when their excited electrons fall to lower energy levels, so all substances absorb characteristic photons when ground state electrons jump to higher energy levels. Thus, the range of photons absorbed by a given element or compound is unique for that element and is called its **absorption spectrum.** The instrument often used to identify the photons absorbed by a given substance is a **spectrophotometer.** Photons which are not absorbed are said to be transmitted.

The Spectronic 20 is the instrument commonly used in academic laboratories. The light source is a tungsten lamp which produces light over a specific range of wavelengths. A "tuner" or "filter" (usually a grating or a prism) selects a narrow band of wavelengths produced by the light source and sends it at a given intensity through the sample. A sample of a substance in a special glass sample holder called a cuvette, will absorb some of the light and allow some of the light to pass through (be transmitted). A detector (usually a phototube) measures the light beam which is transmitted and converts it to an electric current which will then move a needle on the dial or generate a digital readout. The dial or readout can be calibrated as absorbance (range from 0 to 1.5) or percent transmittance (range from 0 to 100%).

PROCEDURE

Wear protective glasses.

 Dispose of all solutions in waste containers provided.

A. Using a spring to generate and observe a transverse wave and its properties

Find a space on the floor where there is plenty of room to stretch a long (about 1.75 m) spring to about 4 m. A long hallway works very well for this. Two students should sit or kneel on the floor with the spring flat on the floor between them, each holding opposite ends securely. If the spring has a ring on each end, do **not** hold the spring by these rings which sometimes slip off and allow a sudden recoil of the spring. Place a short strip of masking tape on the floor at each end of the spring as shown in Figure 16.3. This facilitates keeping both ends of the spring at the same positions on the floor as the waves are generated.

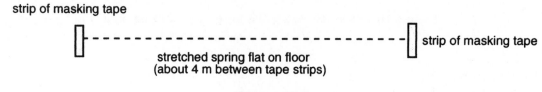

Figure 16.3

One of the two students should now briskly displace the spring (Figure 16.4) using a flip of the wrist while holding their end on the floor in the taped position. The other end of the spring does not move at all.

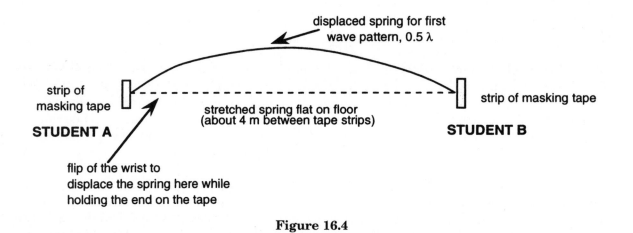

Figure 16.4

Patiently experiment with manipulating the spring to generate the first wave pattern (shown in Figure 16.4) on the floor and then practice generating the second, third, and fourth wave patterns shown on the report form. Several students standing around the stretched spring can evaluate the patterns from a higher vantage and assist student A to know when the displacement and its frequency will best achieve the desired patterns. The first pattern (with a long wavelength) requires the least energy from the wrist of student A. As the wavelengths shorten, the energy required from the wrist of student A increases. Once the team understands what and how the spring manipulations are to be done, proceed as follows.

1. Measure and record the distance between the two pieces of tape in centimeters.

2. Students A and B should begin to generate the first wave form with a wavelength of 0.5 λ. When the wave form is correct, a third student will start counting the vibrations and timing (a stopwatch is best but a watch or clock with a second hand will also work). After 50 vibrations, note and record the elapsed time.

3. Repeat this procedure for waves with 1 λ, 1.5 λ, and 2 λ as shown in the second column on the report form (Part A2). Remove the tape from the floor.

4. Calculate and record the frequency of each wave pattern on the report form. To find frequency, ν:

$$\nu = \frac{\text{no. of waves}}{\text{elapsed time}} = \frac{50 \text{ waves}}{20 \text{ seconds}} = 2.5 \text{ waves/s (or 2.5 cycles/s)}$$

5. Calculate and record the wavelength, λ, for each wave generated. For example, with 400 cm between tapes, the calculation for the wave pattern shown in Figure 16.5 is:

$$\lambda = \frac{400 \text{ cm}}{1.0 \text{ wave}} = 400 \text{ cm/wave}$$

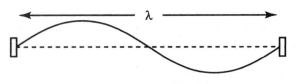

Figure 16.5

6. Use the graph paper provided in the report form to plot frequency vs. wavelength for the waves generated by the spring. In this experiment, the wavelength was the variable being determined by the experimenters. Follow the guidelines in Study Aid 3 for the completion of this graph.

B.   Emission Spectra

1. **Examining the continuous spectrum:**   Take a hand-held spectroscope and aim the end with the slit at an incandescent light. Put the viewing end (with the circular eyepiece that has a diffraction grating over it) up to your eye. To adjust the spectroscope, rotate the entire unit until the spectrum appears vertically on the side walls of the tube. Then, turn the slit portion until the widest spectral band is attained. The bands of color, merging smoothly into each other, are called a **continuous spectrum.** Make a sketch of the continuous spectrum on your report form using colored pencils.

2. **Examining bright-line spectra:**   A **bright-line spectrum** is produced by hot gases of low density when electrons, excited by an electrical current, fall back to lower energy levels and emit photons. The spectrum which results has bright lines which correspond to the energy level transitions separated by dark spaces. The bright lines are determined by the kinds of atoms present in the gases and the amount of energy supplied. Each gas emits its own unique bright-line spectrum. Use the hand-held spectroscope to examine the bright line spectrum of a fluorescent light. Make a sketch of the bright-line spectrum for the fluorescent light in the space provided on your report form.

⚠️ Set up in the laboratory are spectrum tubes for hydrogen and neon. **Do NOT try to adjust these tubes without unplugging the power supply.** Observe the bright line spectrum for each of these gases using your spectroscope. Sometimes it is necessary to decrease the room light to see the lines from the spectrum tubes clearly. Sketch the bright-line spectrum for each gas on your report form.

C. Absorption Spectrum

1. The Spectrophotometer

Plug in the power line and switch the instrument on by turning the control knob clockwise past the click. Allow the instrument to warm-up for twenty minutes before making any measurements.

2. Measurement of the Absorption Spectra for Colored Solutions

The absorption spectrum for two aqueous solutions with visible color (referred to as the samples) will be measured using a Spectronic 20 instrument. The glass sample holders (cuvettes) in which the solutions are placed for these measurements must be clean and unscratched. Fill one cuvette half full with the solvent used for these solutions (distilled water). Fill the other two cuvettes half full with the nickel nitrate (green) and potassium permanganate (purple) solutions provided. This makes a total of three cuvettes that will be used. Several students can work together to measure the absorption spectrum for these solutions as follows:

a. Be sure the machine has been warmed up for about twenty minutes and that you understand the operation of this instrument before starting. This may require a brief demonstration by the instructor.

b. Select the desired wavelength using the wavelength knob. You will start with a wavelength of 350 nm. Insert the cuvette with distilled water into the sample holder and close the cover. Adjust the light control knob so 100% transmittance is read on the scale. This step is known as **calibrating** the instrument. You have adjusted the instrument so that 100% of the light of this wavelength (350 nm) has passed through the sample.

c. Remove the cuvette of distilled water from the sample holder, replace it with the green nickel nitrate solution and close the cover. Read and record the percent transmittance on the report form.

d. Remove the cuvette with the nickel(II) nitrate solution, replace it with the purple potassium permanganate solution and close the cover. Read and record the percent transmittance as before.

e. Repeat steps b–d using the next wavelength shown on the data table (375 nm). It is necessary to calibrate the instrument with distilled water each time you change the wavelength, so repeat step b at each wavelength.

f. Continue until you have measured the percent transmittance for both solutions at wavelengths increasing by 25 nm up to 700 nm. If the available instrument cannot provide the full range of 350–700 nm wavelengths, ask the instructor how to proceed.

3. Graphing the Absorption Spectra Data

Plot the transmittance vs. wavelength data on the graph paper provided. If available, a computer can be used to plot and print this graph as described in Study Aid 3. If the computer option is used, attach a print-out of the resulting graph to the report form.

a. For this graph, the independent variable is the wavelength; the dependent variable is the percent transmittance.

b. Plotting this data will result in two lines, one for each solution. Design your graph with a figure legend (key) using lines and symbols that make it clear which line corresponds to which solution. It is very helpful to use colored pencils to make or cover the lines.

REPORT FOR EXPERIMENT 16

Electromagnetic Energy and Spectroscopy

A.   Wave Properties

1. Length of stretched spring _____ cm

2. Complete the table below for each of the waves generated by your group.

Form of Wave	Length of Wave Form	No. of cycles	Time (s)	Frequency, cycles/s	Wavelength, cm/wave
⌢	0.5 λ	50			
∿	1 λ	50			
∿	1.5 λ	50			
∿	2 λ	50			

3. What does the spring have to do with electromagnetic energy?

4. Plot a graph of frequency vs. wavelength for the data produced by the spring. Label the graph with a suitable title, determine an appropriate scale for each axis, and label with units that match the data. See Study Aid 3 for additional help if necessary.

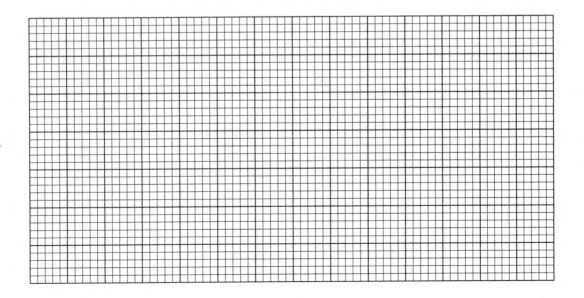

B. Emission Spectra

1. Use colored pencils and sketch the spectrum observed with the spectroscope for each of the light sources observed.

Light Source	Emission Spectrum Observed
Incandescent Bulb	
Fluorescent Bulb	
Hydrogen Gas	
Neon Gas	

2. a. Why must the hydrogen vapor lamp be turned on before it gives off light?

   b. Why is the spectroscope necessary to observe the hydrogen spectrum?

3. Use the colored pencils to color the arrows on the electron transition diagram for H_2 so they correspond to the colors of the visible H_2 spectral lines. Refer to the atomic spectrum chart to match the color to corresponding wavelength then decide the corresponding energy content. Remember that *the length of the arrow is a function of the energy content of the photon released and NOT its wavelength.*

4. Why can we see only 3 or 4 lines in the spectroscope when there are many more arrows in the hydrogen electron transition diagram?

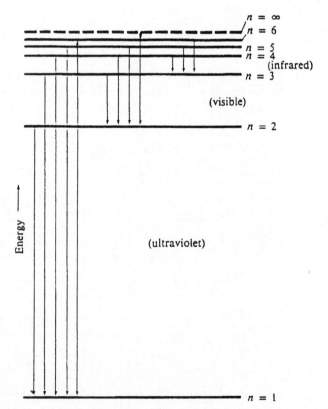

C.   Absorption Spectra for colored solutions

1. Record the percent transmittance data measured from the spectrophotometer for each of the solutions shown on the table below.

Percent transmittance

Wavelength, nm	Ni(NO$_3$)$_2$ (green)	KMnO$_4$ (purple)
350		
375		
400		
425		
450		
475		
500		
525		
550		
575		
600		
625		
650		
675		
700		

2. Graph these data as described in the procedure using either the graph paper provided or a computer if available.

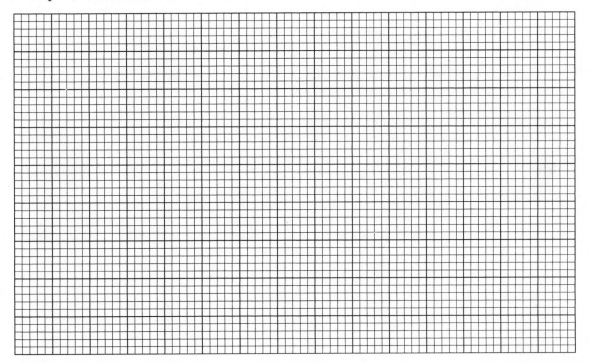

QUESTIONS AND PROBLEMS

1. Draw a diagram on the line below which shows 2.5 transverse waves. Measure the line
    and calculate the wavelength of a single wave in centimeters.

    _____

2. If your diagram represents a wave being generated with a spring like the one used in the
    experiment, and it took 25 seconds to generate 60 of these wave forms, what is the fre-
    quency of the wave? Show calculations.

3. What is the difference between an emission spectrum and an absorption spectrum?

4. What is the relationship between percent transmittance and absorption?

5. What is the relationship between percent transmittance and the color of the solutions?

6. Where do the photons that are absorbed go when they are absorbed by the solution?

EXPERIMENT 17

Lewis Structures and Molecular Models

MATERIALS AND EQUIPMENT

Special equipment: Ball-and-stick molecular model sets

DISCUSSION

Molecules are stable groups of covalently bonded atoms, usually nonmetallic atoms. Chemists study models of molecules to learn more about their bonds, the spatial relationships between atoms and the shapes of molecules. Using models helps us to predict molecular structure.

A. Valence Electrons

Every atom has a nucleus surrounded by electrons which are held within a region of space by the attractive force of the positive protons in the nucleus. The electrons in the outermost energy level of an atom are called valence electrons. The **valence electrons** are involved in bonding atoms together to form compounds. For the representative elements, the number of valence electrons in the outermost energy level is the same as their group number in the periodic table (Groups 1A–7A). For example, sulfur in Group 6A has six valence electrons and potassium in Group 1A has one valence electron.

B. Lewis Structures

Lewis electron dot structures are a useful device for keeping track of valence electrons for the representative elements. In this notation, the nucleus and core electrons are represented by the atomic symbol and the valence electrons are represented by dots around the symbol. Although there are exceptions, Lewis structures emphasize an octet of electrons arranged in the noble gas configuration, ns^2np^6. Lewis structures can be drawn for individual atoms, monatomic ions, molecules, and polyatomic ions.

1. **Atoms and Monatomic Ions:** A Lewis structure for an atom shows its symbol surrounded by dots to represent its valence electrons. Monatomic ions form when an atom loses or gains electrons to achieve a noble gas electron configuration. The Lewis structure for a monatomic ion is enclosed by brackets with the charge of the ion shown. The symbol is surrounded by the valence electrons with the number adjusted for the electrons lost or gained when the ion is formed. This is the basis of ionic bond formation which is not included in this experiment.

Examples: sulfur atom   sulfide ion   potassium atom   potassium ion

$:\!\ddot{S}\!\cdot$ $[:\!\ddot{\ddot{S}}\!:]^{2-}$ $K\cdot$ $[K]^+$

2. **Molecules and Polyatomic Ions:** Lewis structures for molecules and polyatomic ions emphasize the principle that atoms in covalently bonded groups achieve the noble gas configuration, ns^2np^6. Since all noble gases except helium have eight valence electrons, this is

often called the octet rule. Although many molecules and ions have structures which support the octet rule, it is only a guideline. There are many exceptions. One major exception is the hydrogen atom which can covalently bond with only one atom and share a total of two electrons to form a noble gas configuration like helium. All of the examples in this experiment follow the octet rule except hydrogen.

A Lewis structure for covalently bonded atoms is a two-dimensional model in which one pair of shared electrons between two atoms is a single covalent bond represented by a short line; unshared or lone pairs of electrons are shown as dots. Sometimes two pairs of electrons are shared between two atoms forming a double bond and are represented by two short lines. It is even possible for two atoms to share three pairs of electrons forming a triple bond, represented by three short lines. For a polyatomic ion, the rules are the same except that the group of atoms is enclosed in brackets and the overall charge of the ion is shown. For example:

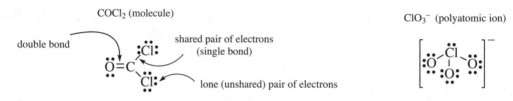

The rules for writing Lewis structures for molecules and polyatomic ions will be provided in the procedure section so you can use your Lewis structures to build three-dimensional models.

C. Molecular Model Building

The three-dimensional structure of a molecule is difficult to visualize from a two-dimensional Lewis structure. Therefore, in this experiment, a ball-and-stick model kit (molecular "tinker toys") is used to build models so the common geometric patterns into which atoms are arranged can be seen. Each model that is constructed must be checked by the instructor and described by its geometry and its bond angles on the report form.

D. Molecular Geometry

Atoms in a molecule or polyatomic ion are arranged into geometric patterns that allow their electron pairs to get as far away from each other as possible (which minimizes the repulsive forces between them). The theory underlying this molecular model is known as the valence shell electron pair repulsion **(VSEPR) theory.** All of the geometric structures in this experiment fall into the following patterns:

1. **Tetrahedral:** four pairs of shared electrons (no pairs of lone (unshared) electrons) around a central atom.

2. **Trigonal pyramidal:** three pairs of shared electrons and one pair of unshared electrons around a central atom.

3. **Trigonal planar:** three groups of shared electrons around a central atom. Two of these groups are single bonds and one group is a double bond made up of two pairs of shared electrons. There are no unshared electrons around the central atom.

4. **Bent:** two groups of shared electrons (in single or double bonds) and one or two pairs of unshared electrons around a central atom.

5. **Linear:** two groups of shared electrons, usually double bonds with two shared electron pairs between two atoms, and no unshared electrons around a central atom. When there are only two atoms in a molecule or ion, and there is no central atom (HBr, for example), the geometry is also linear. These patterns are described more extensively in Section E, which follows.

NOTE: There are other electron arrangements and molecular geometries. Since they do not follow the octet rule, they are not included in this experiment.

E.   Bond Angles

Bond angles always refer to the angle formed between two end atoms with respect to the central atom. If there is no central atom, there is no bond angle.

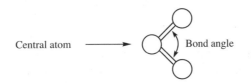

The size of the angle depends mainly on the repulsive forces of the electrons around the central atom. The molecular model kits are designed so that these angles can be determined when sticks representing electron pairs are inserted into pre-drilled holes.

1.   Bond angles for atoms bonded to a central atom without unshared electrons on the central atom.

a.   For four pairs of shared electrons around a central atom (tetrahedral geometry) the angle between the bonds is approximately **109.5°**.

b.   For three atoms bonded to a central atom, (trigonal planar) the angle is **120°**. The shared electron pairs can be arranged in single or double bonds.

c.   For two atoms bonded to a central atom (linear) the angle is **180°**. The shared electrons are usually arranged in double bonds.

d.   Linear diatomic molecules or ions with no central atom do not have a bond angle.

2.   Bond angles for atoms bonded to a central atom **with** unshared electrons on the central atom.

When some of the valence electrons around a central atom are unshared, the VSEPR theory can be used to predict changes in spatial arrangements. An unshared pair of electrons on the central atom has a strong influence on the shape of the molecule. It reduces the angle of bonding pairs by squeezing them toward each other.

For example:

tetrahedral	trigonal pyramidal	bent	bent
No unshared electrons	1 unshared electron pair	2 unshared lone pairs	1 unshared pair
4 pairs shared electrons	3 pairs shared electrons	2 pairs shared electrons	2 groups shared electrons
CH_4	NH_3	H_2O	

109.5° H—C—H 109.5° / 109.5°
with H below C

$H—N(..)—H$ 107° H 107°

$H—O(..)(..)—H$ 105°

$H—S(..)=O$ 119°

repulsive force on shared e⁻ increases, which pushes down on H atoms

repulsive force on shared e⁻ increases, which pushes down on peripheral atoms

F. Bond Polarity

Electrons shared by two atoms are influenced by the positive attractive forces of both atomic nuclei. For like atoms, these forces are equal. For example, in diatomic molecules such as H_2 or Cl_2 the bonded atoms have exactly the same electronegativity (affinity for the bonding electrons). Electronegativity values for most of the elements have been assigned.

Electronegativity Table

1 H 2.1																		2 He
3 Li 1.0	4 Be 1.5											5 B 2.0	6 C 2.5	7 N 3.0	8 O 3.5	9 F 4.0	10 Ne
11 Na 0.9	12 Mg 1.2											13 Al 1.5	14 Si 1.8	15 P 2.1	16 S 2.5	17 Cl 3.0	18 Ar
19 K 0.8	20 Ca 1.0	21 Sc 1.3	22 Ti 1.4	23 V 1.6	24 Cr 1.6	25 Mn 1.5	26 Fe 1.8	27 Co 1.8	28 Ni 1.8	29 Cu 1.9	30 Zn 1.6	31 Ga 1.6	32 Ge 1.8	33 As 2.0	34 Se 2.4	35 Br 2.8	36 Kr
37 Rb 0.8	38 Sr 1.0	39 Y 1.2	40 Zr 1.4	41 Nb 1.6	42 Mo 1.8	43 Tc 1.9	44 Ru 2.2	45 Rh 2.2	46 Pd 2.2	47 Ag 1.9	48 Cd 1.7	49 In 1.7	50 Sn 1.8	51 Sb 1.9	52 Te 2.1	53 I 2.5	54 Xe
55 Cs 0.7	56 Ba 0.9	57–71 La–Lu 1.1–1.2	72 Hf 1.3	73 Ta 1.5	74 W 1.7	75 Re 1.9	76 Os 2.2	77 Ir 2.2	78 Pt 2.2	79 Au 2.4	80 Hg 1.9	81 Tl 1.8	82 Pb 1.8	83 Bi 1.9	84 Po 2.0	85 At 2.2	86 Rn
87 Fr 0.7	88 Ra 0.9	89–103 Ac–Lr 1.1–1.7	104 Rf —	105 Db —	106 Sg —	107 Bh —	108 Hs —	109 Mt —	110 Ds —	111 Rg —							

Key: 9 — Atomic number, F — Symbol, 4.0 — Electronegativity

* The electronegativity value is given below the symbol of each element.

In general, electronegativity increases as we move across a period and up a group on the periodic table. Identical atoms with identical attractions for their shared electron pairs form **nonpolar covalent bonds.** Unlike atoms exert unequal attractions for their shared electrons and form **polar covalent bonds.**

Electronegativity is used to determine the direction of bond polarity which can be indicated in the Lewis structure by replacing the short line for the bond with a modified arrow ($\longmapsto$) pointed towards the more electronegative atom. For example, nitrogen and hydrogen have electronegativity values of 3.0 and 2.1, respectively. The N—H bond is thus represented as

$N \longleftrightarrow H$ with the arrow directed toward the more electronegative nitrogen atom. Then, the Lewis structure can be redrawn with arrows replacing the dashes as shown for NH_4^+ and NH_3.

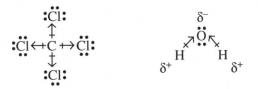

G. Molecular Dipoles

When there are several polar covalent bonds within a molecule or a polyatomic ion such as in NH_3 and NH_4^+ the polar effect of these bonds around a central atom can be cancelled if they are arranged **symmetrically** as shown in CCl_4 below. On the other hand, if the arrangement of the polar bonds is asymmetrical, as in the bent water molecule, H_2O, the resulting molecule has a definite positive end and oppositely charged negative end, and the molecule is called a dipole. In water, the H atoms have a partial positive charge, δ^+, and the O atom has a partial negative charge, δ^-. The symmetry, or lack of symmetry of molecules and polyatomic ions, can generally be seen in the three-dimensional model.

PROCEDURE

Follow steps **A–G** for each of the molecules or polyatomic ions listed on the report form. Refer back to the previous discussion, organized into corresponding sections A–G, for help with each step if necessary.

A. Number of Valence Electrons in a Molecule or Polyatomic Ion

Use a periodic table to determine the number of valence electrons for each group of atoms in the first column of the report form.

    example: SiF_4 Si is in Group 4A, it has 4 valence electrons
                          F is in Group 7A, it has 7 valence electrons

    Total valence electrons is $4 + 4(7) = 32$ electrons

    If the group is a polyatomic ion, total the electrons as above, then add one electron for each negative charge or subtract one electron for each positive charge.

    example: CO_3^{2-} C is in group 4A, it has 4 valence electrons
                          O is in Group 6A, it has 6 valence electrons
                          Ion has a -2 charge, add 2 electrons

Total valence electrons is $4 + 3(6) + 2 = 24$ electrons

B. Lewis Structures for Molecules and Polyatomic Ions

Use the following rules to show the two-dimensional Lewis structure for each molecule or polyatomic ion. Put your structure in the space provided. Use a *sharp* pencil and be as neat as possible.

1. Write down the skeletal arrangement of the atoms and connect them with a single covalent bond (a short line). We want to keep the rules at a minimum for this step, but we also want to avoid arrangements which will later prove incorrect. Useful guidelines are

    a. carbon is usually a central atom or forms bonds with itself; if carbon is absent, the central atom is usually the least electronegative atom in the group;

    b. hydrogen, which has only one valence electron, can form only one covalent bond and is never a central atom;

    c. oxygen atoms are not normally bonded to each other except in peroxides, and oxygen atoms normally have a maximum of two covalent bonds (two single bonds or one double bond).

Using these guidelines, skeletal arrangements for SiF_4 and CO_3^{2-} are

2. Subtract two electrons from the total valence electrons for each single bond used in the skeletal arrangement. This calculation gives the net number of electrons available for completing the electron structure. In the examples above, there are 4 and 3 single bonds, respectively. With 2 e$^-$ per bond the calculation is

SiF_4: $32\,e^- - 4(2\,e^-) = 24\,e^-$ left to be assigned to the molecule

CO_3^{2-}: $24\,e^- - 3(2\,e^-) = 18\,e^-$ left to be assigned to the polyatomic ion

3. Distribute these remaining electrons as pairs of dots around each atom (except hydrogen) to give each atom a total of eight electrons around it. If there are not enough electrons available, move on to step 4.

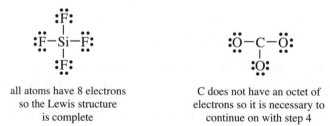

all atoms have 8 electrons so the Lewis structure is complete

C does not have an octet of electrons so it is necessary to continue on with step 4

4. Check each Lewis structure to determine if every atom except hydrogen has an octet of electrons. If there are not enough electrons to give each of these atoms eight electrons, change single bonds between atoms to double or triple bonds by shifting unshared pairs of electrons as needed. A double bond counts as $4\,e^-$ for each atom to which it is bonded.

For CO_3^{2-}, shift $2\,e^-$ from one of the O atoms and place it between C and that O.

Now, all the atoms have $8e^-$ around them. (Don't forget the $^{2-}$)

C.   Model Building

1. Use the balls and sticks from the kit provided to build a 3-dimensional model of the molecule or polyatomic ion for each Lewis structure in the report form.

  a. Use a ball with 4 holes for the central atom.

  b. Use inflexible sticks for single bonds.

  c. Use flexible connectors for double or triple bonds.

  d. Use inflexible sticks for lone pairs around the central atom only.

2. **Leave the model together until it is checked by the instructor.** If you have to wait for someone to check your model, start building the next model on the list. If you complete each structure so fast that you run out of components before someone checks your models, work on other parts of the experiment.

D.   Molecular Geometry

Look at your model from all angles and compare its structure to the description in the discussion (Section D). Then identify its molecular geometry from the following list and write the name of the geometric pattern on the report form in column D.

1. tetrahedral

2. trigonal pyramidal

3. trigonal planar

4. bent

5. linear

E.   Central Bond Angles

Fill in column E with the bond angles between the central atom and all atoms attached to it. Review the discussion (Section E) to find the value of the angles associated with each geometric form. For molecules with more than one central atom, give bond angles for each. For molecules without a central atom and hence no bond angle, write *no central atom*.

F.   Bond Polarity

Bond polarity can be determined by looking up the electronegativity values for both atoms in the Electronegativity table. In the F column of the report form, draw the symbols for both atoms involved in a bond and connect them with an arrow pointing toward the more electronegative atom. If there are several identical bonds it is only necessary to draw one. Use the following as examples.

$$N \leftarrow H \qquad\qquad S \rightarrow O$$

G. Molecular Dipoles

Look at the model and evaluate its symmetry. Decide if the polar bonds within it cancel each other around the central atom resulting in a nonpolar molecule or if they do not cancel one another and result in a dipole. Some examples:

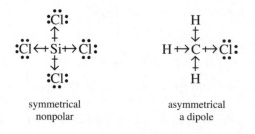

symmetrical
nonpolar

asymmetrical
a dipole

Remember, it is also possible for all the polar bonds within a polyatomic ion to cancel each other so the resultant effect is nonpolar even though the group as a whole has a net charge.

$$\left[\begin{array}{c} O \\ \diagdown \\ \diagup C \Rrightarrow O \\ O \end{array} \right]^{2-}$$

symmetrical
not a dipole

REPORT FOR EXPERIMENT 17

Lewis Structures and Molecular Models

For each of the following molecules or polyatomic ions, fill out columns A through G using the instructions provided in the procedure section. These instructions are summarized briefly below.

A. Calculate the total number of valence electrons in each formula.

B. Draw a Lewis structure for the molecule or ion which satisfies the rules provided in the procedure.

C. Build a model of the molecule and have it checked by the instructor.

D. Use your model to determine the molecular geometry for this molecule (don't try to guess the geometry without the model): tetrahedral, trigonal pyramidal, trigonal planar, bent, linear

E. Determine the bond angle between the central atom and the atoms bonded to it. If there are only two atoms in the structure write "no central atom" in the space provided.

F. Use the electronegativity table to determine the electronegativity of the bonded atoms.
If the bonds are polar, indicate this with a modified arrow ($\longmapsto$) pointing to the more electronegative element.
If the bonds are nonpolar, indicate this with a short line (—).
If there are two or more different atoms bonded to the central atom, include each bond.

G. Use your model and your knowledge of the bond polarity to determine if the molecule as a whole is nonpolar or a dipole. If it is polar, write *dipole* in G. If it is not, write *nonpolar*.

Molecule or Polyatomic Ion	No. of Valence Electrons	Lewis Structure	Molecular Geometry	Bond Angles	Bond Polarity	Molecular Dipole or Nonpolar
	A	B	C	D	E	F	G
CH_4							
CS_2							

Molecule or Polyatomic Ion	A — No. of Valence Electrons	B — Lewis Structure	C	D — Molecular Geometry	E — Bond Angles	F — Bond Polarity	G — Molecular Dipole or Nonpolar
H_2S							
N_2							
SO_4^{2-}							
H_3O^+							
CH_3Cl							
C_2H_6							
C_2H_4							

	A	B	C	D	E	F	G
No. of Molecule or Polyatomic Ion	Valence Electrons	Lewis Structure		Molecular Geometry	Bond Angles	Bond Polarity	Molecular Dipole or Nonpolar
$C_2H_2Cl_2$		*					
SO_3^{2-}							
CH_2O							
OF_2							
NO_2^-							
O_2							
NO_3^-		**					

*More than one possible Lewis structure can be drawn. See questions 1, 2.
**More than one possible Lewis structure can be drawn. See question 3.

QUESTIONS

1. There are three acceptable Lewis structures for $C_2H_2Cl_2$ (*) and you have drawn one of them on the report form. Draw the other two structures and indicate whether each one is nonpolar or a dipole.

2. Explain why one of the three structures for $C_2H_2Cl_2$ is nonpolar and the other two are molecular dipoles.

3. There are three Lewis structures for $[NO_3]^-$ (**). Draw the two structures which are not on the report form. Compare the molecular polarity of the three structures.

EXPERIMENT 18

Boyle's Law

MATERIALS AND EQUIPMENT

Special equipment: Elasticity of Gases Kit (or Simple Form Boyle's Law Apparatus); tube of silicone grease, barometer, slotted 0.5 and 1.0 kg masses, assorted bricks, and a vernier caliper.

DISCUSSION

Matter in the gaseous state has neither a definite volume nor a definite shape. Therefore, a confined sample of gas will take the shape of its container and depending on conditions, its volume can increase or decrease. In this experiment, air is the gas and its container is a large plastic syringe as shown in Figure 18.1. There are two conditions that can change the volume of this confined gas sample: pressure and temperature. We will keep the temperature constant (at room conditions) and examine the quantitative relationship between the volume and pressure by adding masses to the platform on the syringe. As these masses are added, the barrel will move relative to the piston and the volume of the confined gas sample will decrease as pressure, P_{gas}, increases.

This relationship was first recognized by the British scientist Robert Boyle in 1662 and is known as Boyle's Law: *at constant temperature, the volume of a sample of gas varies inversely with the pressure.* The statement may be symbolized as follows (Equation 1):

$$V \propto \frac{1}{P} \qquad \text{(constant T)} \tag{1}$$

$$V = \mathbf{k} \times \frac{1}{P} \qquad \text{(constant T)} \tag{2}$$

$$PV = \mathbf{k} \qquad \text{(constant T)} \tag{3}$$

where $\mathbf{k}$ is a constant that depends on the mass and the temperature of the gas.

Equation (2) emphasizes the inverse relationship between pressure and volume which can be stated in simpler language: As the pressure on a gas is increased, the volume is decreased and vice versa.

Equation (3) is obtained by rearranging equation (2) and states that, at constant temperature, the product of the pressure and volume of a given mass of gas is *constant*. From equation (3) it follows that $P_1V_1 = k = P_2V_2$, therefore:

$$P_1V_1 = P_2V_2 \qquad \text{(constant T)} \tag{4}$$

where P_1V_1 is the pressure-volume product at one set of conditions and P_2V_2 is the product at a second set of conditions. Solving equation (4) for V_2 gives

$$V_2 = \frac{V_1P_1}{P_2} \qquad \text{(constant T)} \tag{5}$$

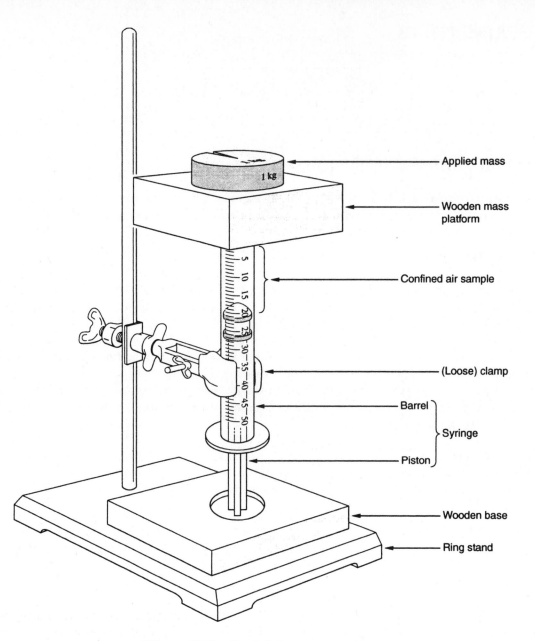

Figure 18.1   Boyle's law apparatus

This equation is commonly used in Boyle's law calculations. To calculate a new volume (V_2), the initial volume is multiplied by the ratio of the initial pressure over the final pressure (P_1/P_2).

In this experiment the pressure and volume of a gas are measured. The data are plotted as volume vs. pressure to obtain a curve typical of an inverse relationship. The pressure-volume product (PV) is calculated for each set of data. The constancy of these PV products proves the validity of equation (3) and thus the validity of Boyle's law.

The apparatus is essentially a large syringe mounted on a wooden base. The barrel of the syringe is the container for the gas with a pressure P_{gas}. This pressure is equal to the

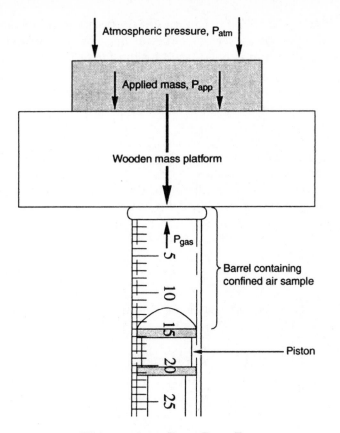

Figure 18.2 P_{atm}, P_{app}, P_{gas}

pressure of the atmosphere, P_{atm}, pushing down on the top of the barrel and any applied pressure added to the platform, P_{app}. This is illustrated in Figure 18.2 and summarized as

$$P_{gas} = P_{atm} + P_{app} \tag{6}$$

Gas pressure can be measured in various units. Atmospheric pressure P_{atm} is usually measured with a barometer calibrated in mm Hg or in. Hg. The applied pressure (P_{app}) is measured in units of mass per surface area.

$$P_{app} = \frac{mass}{surface\,area} \tag{7}$$

One commonly used pressure unit is lb/in.^{2} (read as "pounds per square inch" and sometimes abbreviated as "psi"). It is measured by calculating the weight of the masses in pounds and dividing by the cross-sectional area of the barrel in square inches. Different equivalent units of pressure are related as follows:

$$1\,atm = 760\,mmHg = 760\,torr = 14.7\,lb/in.^2\,(psi)$$

For a barometer reading of 747 mmHg, the P_{atm} in lb/in.^{2} is calculated as follows:

$$(747\,mmHg)\left(\frac{14.7\,lb/in.^2}{760\,mmHg}\right) = 14.4\,lb/in.^2$$

In the procedure, the masses added to the weight platform are kilogram-slotted masses and bricks. Their force is applied to the cross-sectional area of the syringe. The cross-sectional area of the syringe is circular and is calculated by the formula for area (A) of a circle:

$$A = \pi r^2 \qquad \text{(where } r = d/2\text{)} \qquad (8)$$

r = radius and d = diameter

$\pi = 3.14$

To calculate the area, the inside diameter of the syringe is measured using a vernier caliper calibrated in either centimeters or inches (or both). If you are unfamiliar with the use of a vernier caliper, your instructor will demonstrate its use. In the example illustrated (Figure 18.3), the inside diameter of the barrel is 32.7 mm or 3.27 cm. Substituting into equation 8, the cross sectional area of the barrel illustrated is calculated as

$$A = (3.14)\left(\frac{3.27\,\text{cm}}{2}\right)^2 = 8.39\,\text{cm}^2$$

With a 2.5 kg mass on the weight platform, the applied pressure, P_{app} is 2.5 kg/8.39 cm^2 which, when converted to lb/in.^{2}, is 4.2 lb/in.^{2}

$$P_{app} = \frac{\text{mass}}{A} = \left(\frac{2.5\,\text{kg}}{8.39\,\text{cm}^2}\right)\left(\frac{2.2\,\text{lb}}{1\,\text{kg}}\right)\left(\frac{2.54\,\text{cm}}{1\,\text{in.}}\right)^2 = 4.2\frac{\text{lb}}{\text{in.}^2}$$

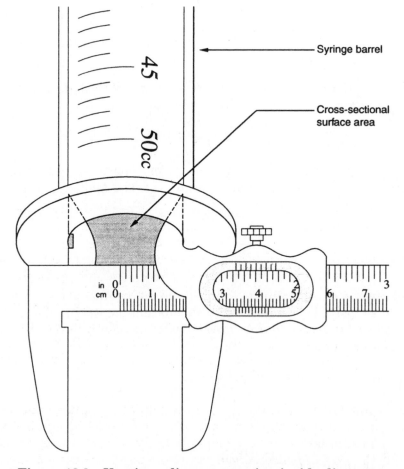

Figure 18.3 Vernier caliper measuring inside diameter

Therefore, the pressure on the gas in the syringe at atmospheric pressure with a 2.5 kg mass on the mass platform is:

$$P_{gas} = P_{atm} + P_{app}$$

$$P_{gas} = 14.4\,lb/in.^2 + 4.2\,lb/in.^2 = 18.6\,lb/in.^2$$

PROCEDURE

Wear protective glasses.

WASTE DISPOSE OF PROPERLY **No waste for disposal in this experiment.**

> Record all measurements on the data tables of the report form and perform calculations as indicated. Every calculated number entered on the report form must have a setup shown clearly in the space provided.

1. Read the barometer and record the atmospheric pressure in mm Hg.

2. Remove the red cover from the nozzle of the syringe and pull the piston all the way out. Measure the inside diameter of the barrel with the vernier caliper. Be careful not to lose the red cap.

3. Lubricate the side wall of the black rubber gasket on the bottom end of the piston with silicone grease.

4. Seat the top end of the piston firmly in the hole in the thin wooden block, the base.

5. Place the barrel over the gasket on the piston and move it to the 35.0 cm³ (cc) mark on the barrel. Put the red cap back on the nozzle of the barrel.

6. Place the yellow plastic cap over the top of the barrel with the cap passing through the center hole (some kits may not have this yellow cover). Put the wooden base of this whole assembly on the ring stand and place a clamp around the barrel but do not tighten it. Turn the barrel so the calibrations are visible with the clamp in place.

7. Put the wooden mass platform (the thick wooden block) over the top of the barrel so the red cap on the nozzle is below the surface of the surrounding wood. Push down and release, allowing the barrel to bounce back up and come to a stop.

8. With no applied mass on the mass platform, record the volume of the air inside the cylinder in the appropriate space on Data Table 2 of the report form (top line of the table, applied mass = 0). The only pressure now exerted on the gas inside the syringe is atmospheric pressure and that of the wooden platform.

9. Weigh one of the 0.5 or 1 kg masses to three significant figures and place it carefully on the mass platform so the syringe does not bend and touch the stabilizing clamp. Push down and release as before. One person should hold the syringe steady while another person carefully adds the mass. A third person can read and record the volume. As the masses are added, the syringe has a tendency to become unsteady and the masses can slide off or

the syringe can break. The ring stand and clamp are supposed to prevent this kind of disaster but teamwork is necessary to make sure the clamp doesn't create more problems by interfering with the movement of the barrel on the piston. Read and record the volume.

10. Repeat Step 9 by adding various combinations of weighed masses and bricks.

When complete, there should be five values recorded in the first and fifth columns of Data Table 2, starting with an applied mass of 0 in the top row.

11. On your report form, complete Data Table 1, 5–10 for the first applied mass. Show the setups for every calculated number and pay attention to significant figures. Complete the calculations for the remaining applied masses.

12. Graph the data using the following guidelines.

   a. The variables in this experiment are the pressure of the confined gas, P_{gas}, and the volume of the gas under each pressure condition. The independent variable is the variable which the experimenter controlled. This variable is plotted on the x-axis. The dependent variable changes in response to changes made in the independent variable. It is plotted on the y-axis.

   b. If necessary, review the instructions in Study Aid 3 for determining scale values, choosing suitable starting values, numbering the major increments, and writing axes labels and a title.

   c. Plot each PV point and draw a smooth line that best fits the 5 points. This line should be slightly curved and need not pass through each point.

REPORT FOR EXPERIMENT 18

Boyle's Law

Data Table 1   Show the setup used for every calculated number. Include all units in dimensional analysis setups.

1. Atmospheric pressure   _____ mm Hg   _____ lb/in.^{2}

2. Inside diameter of syringe   _____ cm   _____ in.

3. Inside radius   _____ in.

4. Area of syringe (cross-section)   _____ in.^{2}

Record measurements and show your calculations for the first applied mass in Data Table 2.

5. First applied mass   _____ kg

6. Applied mass   _____ lb

7. Applied pressure   _____ lb/in.^{2}

8. Total pressure   _____ lb/in.^{2}

9. Volume of air   _____ cm^3

10. Pressure-Volume Product, PV   _____ lb cm^3/in.^{2}

Data Table 2   Complete the table below for 0 applied mass and 4 additional masses. Calculations for the first applied mass should be shown above in 5–10.

Applied Mass, kg	Applied Mass, lb	Applied Pressure, P_{app} (lb/in.^{2})	Total Pressure $P_{gas} = P_{app} + P_{app}$ (lb/in.^{2})	Volume of Air, V (cm^3)	Pressure-Volume Product, PV (lb cm^3/in.^{2})
0	0	0			

Average PV _____

QUESTIONS AND PROBLEMS

1. What is the independent variable in this experiment? Explain your choice.

2. Why must the temperature be constant during this experiment?

3. What part of the tabulations in the data tables proves Boyle's Law? How?

4. Plot Total Pressure vs. Volume of Air for the five values in Data Table 2. Use the graph paper provided or attach a computer-generated graph to your report form. If necessary, refer to Study Aid 3 for instructions on how to complete either type of graph.

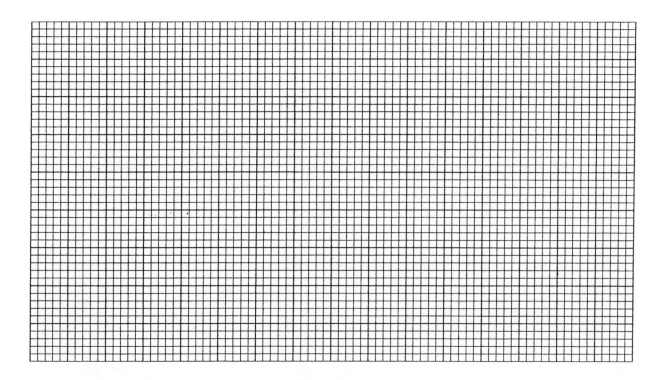

5. What is the relationship between gas pressure and volume shown by the data in your graph?

6. If you repeated this experiment at a lower temperature (for example in a walk-in refrigerator), how would the P vs. V curve obtained differ from the curve on your graph?

7. Given the following data: volume of air without any applied pressure, 25.0 cm^3; inside diameter of the syringe, 3.20 cm; barometric pressure, 630 mm Hg. Show setups and answers for each of the following problems based on this data.

a. What is the cross-sectional area of the syringe in in.^{2}?

_____

b. What is the barometric pressure in lb/in.^{2}?

_____

c. A brick with a mass of 6.00 lb is placed on the barrel of the syringe. What is the total pressure, P_{gas}, of the gas in the syringe after adding the brick?

_____

d. What is the change in volume after adding the brick?

_____

EXPERIMENT 19

Charles' Law

MATERIALS AND EQUIPMENT

125 mL Erlenmeyer flask, one-hole rubber stopper, glass and rubber tubing, pneumatic trough, thermometer, screw clamp.

DISCUSSION

The quantitative relationship between the volume and the absolute temperature of a gas is summarized in Charles' law. This law states: at constant pressure, the volume of a particular sample of gas is directly proportional to the absolute temperature.

Charles' law may be expressed mathematically:

$$V \propto T \qquad \text{(constant pressure)} \qquad (1)$$

$$V = kT \quad \text{or} \quad \frac{V}{T} = k \qquad \text{(constant pressure)} \qquad (2)$$

where V is volume, T is Kelvin temperature, and k is a proportionality constant dependent on the number of moles and the pressure of the gas.

If the volume of the same sample of gas is measured at two temperatures, $V_1/T_1 = k$ and $V_2/T_2 = k$, and we may say that

$$\frac{V_1}{T_1} = \frac{V_2}{T_2} \quad \text{or} \quad V_2 = (V_1)\left(\frac{T_2}{T_1}\right) \qquad \text{(constant pressure)} \qquad (3)$$

where V_1 and T_1 represent one set of conditions and V_2 and T_2 a different set of conditions, with pressure the same at both conditions.

Experimental Verification of Charles' Law

This experiment measures the volume of an air sample at two temperatures, a high temperature, T_H, and a low temperature, T_L. The volume of the air sample at the high temperature, (V_H), decreases when the sample is cooled to the low temperature and becomes V_L. All of these measurements are made directly. The experimental data is then used to verify Charles' law by two methods:

1. The experimental volume (V_{exp}) measured at the low temperature is compared to the V_L predicted by Charles' law where

$$V_L(theoretical) = (V_H)\left(\frac{T_L}{T_H}\right)$$

2. The V/T ratios for the air sample measured at both the high and the low temperatures are compared. Charles' law predicts that these ratios will be equal.

$$\frac{V_H}{T_H} = \frac{V_L}{T_L}$$

Pressure Considerations

The relationship between temperature and volume defined by Charles' law is valid only if the pressure is the same when the volume is measured at each temperature. That is not the case in this experiment.

1. The volume, V_H, of air at the higher temperature, T_H, is measured at atmospheric pressure, P_{atm} in a dry Erlenmeyer flask. The air is assumed to be dry and the pressure is obtained from a barometer.

2. The experimental air volume, (V_{exp}) at the lower temperature, T_L, is measured over water. This volume is saturated with water vapor that contributes to the total pressure in the flask. Therefore, the experimental volume must be corrected to the volume of dry air at atmospheric pressure. This is done using Boyle's law as follows:

    a. The partial pressure of the dry air, P_{DA}, is calculated by subtracting the vapor pressure of water from atmospheric pressure:

$$P_{DA} = P_{atm} - P_{H_2O}$$

    b. The volume that this dry air would occupy at P_{atm} is then calculated using the Boyle's law equation:

$$(V_{DA})(P_{atm}) = (V_{exp})(P_{DA})$$
$$(V_{DA}) = \frac{(V_{exp})(P_{DA})}{(P_{atm})}$$

PROCEDURE

Wear protective glasses.

No waste for disposal in this experiment.

> **NOTE:** It is essential that the Erlenmeyer flask and rubber stopper assembly be as dry as possible in order to obtain reproducible results.

Dry a 125 mL Erlenmeyer flask by gently heating the entire outer surface with a burner flame. Care must be used in heating to avoid breaking the flask. If the flask is wet, first wipe the inner and outer surfaces with a towel to remove nearly all the water. Then, holding the flask with a test tube holder, gently heat the entire flask. Avoid placing the flask directly in the flame. Allow to cool.

While the flask is cooling select a 1-hole rubber stopper to fit the flask and insert a 5 cm piece of glass tubing into the stopper so that the end of the tubing is flush with the bottom of the stopper. Attach a 3 cm piece of rubber tubing to the glass tubing (see Figure 19.1). Insert the stopper into the flask and mark (wax pencil) the distance that it is inserted. Clamp the flask so that it is submerged as far as possible in water contained in a 400 mL beaker (without the flask touching the bottom of the beaker) (see Figure 19.2).

Heat the water to boiling. Keep the flask in the gently boiling water for at least 8 minutes to allow the air in the flask to attain the temperature of the boiling water. Add water as needed to maintain the water level in the beaker. Read and record the temperature of the boiling water.

While the flask is still in the boiling water, seal it by clamping the rubber tubing tightly with a screw clamp. Remove the flask from the hot water and submerge it in a pan of cold water, keeping the top down at all times to avoid losing air (see Figure 19.3). Remove the screw clamp, letting the cold water flow into the flask. Keep the flask totally submerged for about 6 minutes to allow the flask and contents to attain the temperature of the water. Read and record the temperature of the water in the pan.

Figure 19.2   Heating the flask (and air) in boiling water

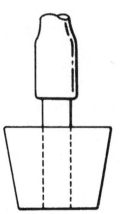

Figure 19.1   Rubber stopper assembly

In order to equalize the pressure inside the flask with that of the atmosphere, bring the water level in the flask to the same level as the water in the pan by raising or lowering the flask (see Figure 19.3). With the water levels equal, pinch the rubber tubing to close the flask. Remove the flask from the water and set it down on the laboratory bench.

Using a graduated cylinder carefully measure and record the volume of water in the flask.

Repeat the entire experiment. Use the same flask and flame dry again; **make sure that the rubber stopper assembly is thoroughly dried inside and outside.**

After the second trial fill the flask to the brim with water and insert the stopper assembly to the mark, letting the glass and rubber fill to the top and overflow. Measure the volume of water in the flask. Since this volume is the total volume of the flask, record it as the volume of air at the higher temperature. Because the same flask is used in both trials, it is necessary to make this measurement only once.

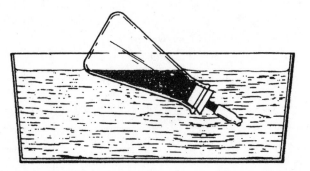

Figure 19.3   Equalizing the pressure in the flask. The water level inside the flask is adjusted to the level of the water in the pan by raising or lowering the flask

REPORT FOR EXPERIMENT 19

Charles' Law

Data Table

	Trial 1	Trial 2
Temperature of boiling water, T_H	_____°C, _____K	_____°C, _____K
Temperature of cold water, T_L	_____°C, _____K	_____°C, _____K
Volume of water collected in flask (decrease in the volume of air due to cooling)		
Volume of air at higher temperature, V_H (volume of flask measured only after Trial 2)		
Volume of wet air at lower temperature (volume of flask less volume of water collected), V_{exp}		
Atmosphere pressure, P_{atm} (barometer reading)		
Vapor pressure of water at lower temperature, P_{H_2O} (see Appendix 6)		

CALCULATIONS: In the spaces below, show calculation setups for Trial 1 only. Show answers for both trials in the boxes

	Trial 1	Trial 2
1. (a)		
1. (b)		
2.		
3.		
4. (a)		
4. (b)		

1. Corrected experimental volume of dry air at the lower temperature calculated from data obtained at the lower temperature.

    (a) Pressure of dry air (P_{DA})

    $$P_{DA} = P_{Atm} - P_{H_2O}$$

    (b) Corrected experimental volume of dry air (lower temperature).

    $$V_{DA} = (V_{exp})\left(\frac{P_{DA}}{P_{Atm}}\right) =$$

2. Predicted volume of dry air at lower temperature V_L calculated by Charles' law from volume at higher temperature (V_H).

    $$V_L = (V_H)\left(\frac{T_L}{T_H}\right)$$

3. Percentage error in verification of Charles' law.

    $$\% \, error = \left(\frac{V_L - V_{DA}}{V_L}\right)(100) =$$

4. Comparison of experimental V/T ratios. (Use dry volumes and absolute temperatures.)

    (a) $\dfrac{V_H}{T_H} =$

    (b) $\dfrac{V_{DA}}{T_L} =$

5. On the graph paper provided, plot the volume-temperature values used in Calculation 4. Temperature data **must be in** °C. Draw a straight line between the two plotted points and extrapolate (extend) the line so that it crosses the temperature axis.

QUESTIONS AND PROBLEMS

1. (a) In the experiment, why are the water levels inside and outside the flask equalized before removing the flask from the cold water?

(b) When the water level is higher inside than outside the flask, is the gas pressure in the flask higher than, lower than, or the same as, the atmospheric pressure? (specify which)

_____

2. A 125 mL sample of dry air at 230°C is cooled to 100°C at constant pressure. What volume will the dry air occupy at 100°C?

_____ mL

3. A 250 mL container of a gas is at 150°C. At what temperature will the gas occupy a volume of 125 mL, the pressure remaining constant?

_____ °C

4. (a) An open flask of air is cooled. Answer the following:

1. Under which conditions, before or after cooling, does the flask contain more gas molecules?

_____

2. Is the pressure in the flask at the lower temperature the same as, greater than, or less than the pressure in the flask before it was cooled?

_____

(b) An open flask of air is heated, stoppered in the heated condition, and then allowed to cool back to room temperature. Answer the following:

1. Does the flask contain the same, more, or fewer gas molecules now compared to before it was heated?

_____

2. Is the volume occupied by the gas in the flask approximately the same, greater, or less than before it was heated?

_____

3. Is the pressure in the flask the same, greater, or less than before the flask was heated?

_____

4. Do any of the above conditions explain why water rushed into the flask at the lower temperature in the experiment? Amplify your answer.

5. On the graph you plotted,

(a) At what temperature does the extrapolated line intersect the x-axis?

_____ °C

(b) At what temperature does Charles' law predict that the extrapolated line should intersect the x-axis?

_____ °C

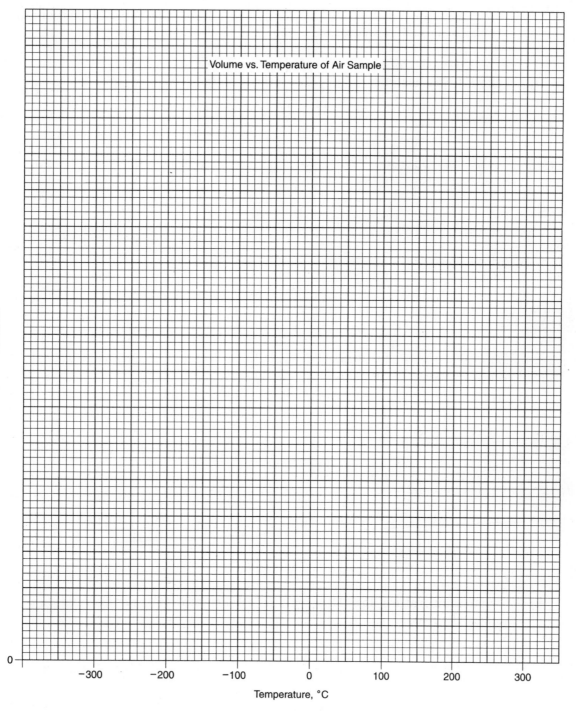

Volume vs. Temperature of Air Sample

Volume, mL

0

−300   −200   −100   0   100   200   300

Temperature, °C

EXPERIMENT 20

Liquids: Vapor Pressure and Boiling Points

MATERIALS AND EQUIPMENT

Liquids: flasks (125 mL) containing acetone [$(CH_3)_2C = O$], methanol (CH_3OH), ethanol (C_2H_5OH), and water. Two test tubes (25×200 mm), boiling chips, fine copper wire (24 gauge), screw clamp, two small pieces rubber tubing, one-gallon metal can with rubber stopper and thermal gloves or towel for Instructor Demonstration (one can per class).

DISCUSSION

The molecules that escape (vaporize) from the surface of a liquid in an open container diffuse into the surroundings. The probability of these escaped molecules returning to condense to a liquid is very small. The volume of the liquid therefore decreases. But the molecules that vaporize from a liquid in a closed container cannot permanently escape. The volume of the liquid therefore does not decrease. An equilibrium is established in which the rate of return to the liquid **(condensation)** is equal to the rate of escape **(evaporation).** The pressure exerted by the vapor in this equilibrium is known as the **vapor pressure** of the liquid. In this experiment you will observe some of the phenomena associated with the vapor pressures of liquids.

Vapor pressure is dependent on two factors: the temperature of the liquid, and the nature of the liquid. As the temperature of the liquid is raised, the vapor pressure increases because the average kinetic energy of the molecules is increased causing more of them to escape from the liquid. When the vapor pressure has increased until it equals the pressure of the atmosphere above it, the liquid boils. The temperature at which the vapor pressure equals exactly one atmosphere (760 torr) is called the **normal boiling point.**

Each liquid has a unique pattern of vapor pressure behavior. Liquids, which have a relatively high vapor pressure, evaporate easily and are said to be **volatile.** In comparing two liquids at a given temperature, the more volatile liquid will have a higher vapor pressure. As a consequence of its higher vapor pressure the more volatile liquid will boil at a lower temperature.

A volatile liquid in an open beaker becomes colder as a result of evaporation. One way of explaining this cooling effect is to assume that the faster moving "hotter" molecules escape from the liquid leaving the slower moving, "cooler" molecules behind.

A.   The Cooling Effect of Evaporation

In the first part of this experiment, the bulb of a thermometer is wrapped with filter paper and wet with a liquid. The evaporation of the liquid causes cooling followed by a drop in temperature on the thermometer. The more volatile the liquid, the faster it will evaporate, and thus, the more the temperature will be lowered.

$$\text{Liquid} + \text{Heat} \xrightarrow{\text{Evaporation}} \text{Vapor}$$

The equation above indicates that the evaporation consumes heat; if no external heat is supplied the temperature of the system will drop as evaporation occurs.

B. The Relationship of Vapor Pressure and Boiling Point

Since the boiling point is the temperature at which the vapor pressure equals the pressure of the atmosphere above a liquid, it follows that the boiling point will be reduced if the pressure above the liquid is reduced.

In this part of the experiment, water is heated to boiling in both the open and stoppered tubes shown in Figure 20.1. The screw clamp attached to the lower end of the delivery tube is left open. The pressure above the water in both tubes is equal to the atmospheric pressure. As the water boils, steam is produced. From the open tube the steam escapes into the atmosphere. From the stoppered tube a mixture of air and steam is expelled through the delivery tube. After a few minutes of boiling, the air has been almost totally expelled and the vapor in the system is almost pure steam (at atmospheric pressure) in the stoppered tube.

Now, heating of both tubes is stopped, and the screw clamp on the stoppered tube system is closed. As the system begins to cool, steam begins to condense on the walls of the stoppered tube and in the delivery tube. The condensing steam reduces the pressure within the system to less than atmospheric pressure. Consequently, the water in the stoppered tube continues to boil for some time because its vapor pressure exceeds the pressure of the atmosphere within the closed system. Because it is boiling, the water in the stoppered tube cools faster than the water in the open tube.

A vapor gives off heat when condensing to a liquid.

$$\text{Vapor} \xrightarrow{\text{Condensation}} \text{Liquid} + \text{Heat}$$

Because steam (vapor) is condensing on the walls of the stoppered tube in the closed system, the walls remain hot for a relatively long time. Since the water in the open tube is not boiling, there is little vapor condensing on the walls of this tube. Therefore the walls cool faster than do the walls of the stoppered tube.

C. Effect of Vapor Pressure Change and Atmospheric Pressure (Instructor Demonstration)

In the last part of this experiment a small volume of water is converted to steam by heating in an unstoppered metal gallon can. When the can is filled with steam it is stoppered and allowed to cool. The condensation of steam inside the can reduces the pressure in the closed can. This is the same phenomenon that occurs in Part B. But a metal can does not have the relative strength of a test tube, so the walls of the can are crushed because of the difference between the atmospheric pressure outside and the reduced pressure inside the can.

PROCEDURE

Wear Protective Glasses

A. Cooling Effect of Evaporation

 PRECAUTIONS:

1. Avoid breathing methanol vapor since it is poisonous.

2. Acetone, methanol, and ethanol are flammable liquids.

1. Wrap the bulb of a thermometer with a half circle of filter paper and fasten the paper securely with fine copper wire. Record the temperature reading of the thermometer (room temperature).

2. A 125 mL flask of each liquid has been prepared for class use. Dip the covered bulb of the thermometer into the flask of acetone. Remove the thermometer from the liquid and suspend it from a ring stand. Note that the temperature begins to decrease. Keep checking the thermometer and when the temperature stabilizes, record the lowest temperature on Data Table A.

3. Repeat the experiment, first using methanol, then ethanol, and then water. Attach a fresh half circle of filter paper each time. **Do the experiment with methanol in the hood.**

 Dispose of filter paper (solvent will be evaporated) in the trash.

B. Relationship of Vapor Pressure and Boiling Point

Work in pairs for this part of the experiment

1. Assemble the apparatus as shown in Figure 20.1. The 25 × 200 mm test tubes **must be scrubbed with detergent and rinsed with distilled water.** Use the precautions described in Experiment 1 when inserting the thermometer and the glass tubing through the rubber stopper. Place two boiling chips and 25 mL of distilled water into each test tube. The stopper must be inserted very securely to avoid leaks and **the screw clamp must be open.**

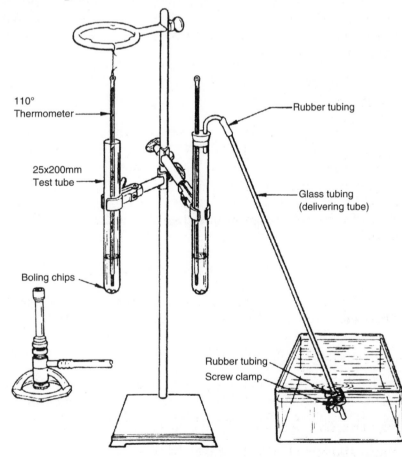

Figure 20.1   Setup for Part B. Relationship of Vapor Pressure and Boiling Point

2. Heat the water in both test tubes to the boiling point by moving a burner flame from one to the other. It is desirable to have the water in both tubes reach the boiling point at nearly the same time. Read the temperature in each tube while the water is boiling. Record these temperatures as first entries in Data Table B. Continue boiling the water in both tubes until there are essentially no bubbles of gas rising to the surface through the water in the trough. When this condition has been reached, heat for an additional minute.

3. Now, in rapid succession

    a. Pinch the rubber tubing shut just below the screw clamp;

    b. Stop heating;

    c. While keeping the tubing pinched shut, close the screw clamp tightly;

    d. When the clamp is tightly closed, note the amount of water in the delivery tube

4. If leaks occur, reassemble the apparatus, checking all the rubber connections, and repeat the experiment (it may be necessary to replace the rubber tubing).

5. Take a series of temperature readings, noting at what temperature boiling stops in each tube. Each time the thermometer **in the closed tube** cools about 5°C, read and record the temperature in both tubes. It is not critical that you have a reading for every 5°C temperature drop, but it is important that both temperatures be read at the same time for each recorded reading.

6. When the water in the open tube has cooled to about 85°C, touch the upper parts of both tubes and note which one is hotter. Also note the condition of the rubber tubing at the top of the closed tube. When the thermometer column is no longer visible because of the rubber stopper, do not try to change the position of the thermometer. Simply wait for the mercury to reappear below the stopper and then continue to take readings.

7. Discontinue taking readings when the temperature in the closed tube falls below 60°C. If the water in either tube is still boiling, record the temperature at which boiling stopped as "below 60°C".

8. When boiling has stopped in the closed tube, observe the amount of water in the delivery tube. Keeping the rubber tubing under water in the trough, open the screw clamp and note what happens.

C. **Effect of Vapor Pressure Change and Atmospheric Pressure (Instructor Demonstration)**

1. Obtain a metal gallon can and a rubber stopper which fits the mouth of the can.

 At this point, the can should NOT be stoppered. Thermal gloves or a towel should be available for handling the can after it is heated.

2. Pour about 50 mL of water into the can, place it on a ring stand, and heat with a burner.

3. After the water has begun to boil and condensing steam has appeared at the mouth of the can, heat for about five more minutes. Stop heating and **remember that the can is very hot** at this point. Using a towel or thermal gloves, immediately stopper the can very tightly.

4. Set the can on the bench and allow it to cool. Observe what happens.

REPORT FOR EXPERIMENT 20

Liquids: Vapor Pressure and Boiling Points

A. Cooling Effect of Evaporation

Room Temperature _____

Data Table A.

Liquid wetting thermometer bulb	Lowest Temperature observed, °C	Normal Boiling Point (See Appendix 7)
Acetone		
Methanol		
Ethanol		
Water		

1. Why does the temperature drop when the thermometer covered with wet filter paper is suspended in the air?

2. On the basis of the observed behavior of these four liquids, what is the relationship between the normal boiling points and effectiveness of liquids in cooling by evaporation?

B. Relationship of Boiling Point and Vapor Pressure

Data Table B. Circle the temperature at which boiling stopped in each tube.

Temperature Readings, °C			
Closed Tube	Open Tube	Closed Tube	Open Tube

1. How much water was in the delivery tube (length of water column above the screw clamp)?

   (a) at the time of closing the screw clamp   _____ cm

   (b) after boiling has stopped   _____ cm

2. When the water had cooled to 85°C in the open tube and you touched the upper part of both test tubes, which was hotter?   _____
   Why?

3. During the cooling, what happened to the rubber tubing at the top of the closed tube? Why?

4. When you opened the screw clamp under water, what happened? Why?

5. Why did the water in the closed test tube continue to boil at lower temperatures than the water in the open test tube?

6. Why did the water in the closed test tube cool faster than the water in the open test tube?

7. Explain what caused the reduced pressure inside the closed test tube during cooling.

8. Why was there more water in the delivery tube after the apparatus cooled (before the clamp was opened) than at the time it was clamped shut.

C.   Effect of Vapor Pressure Change and Atmospheric Pressure

1. (a) What did you observe at the mouth of the can just before it was stoppered?

   (b) What does this indicate about the composition of the gas in the can when the can was stoppered?

2. What happened to the can after heating was stopped and it was stoppered tightly?

3. If the can had been stoppered and heating stopped when the water first began to boil, how might the results have been different? Why?

4. If the can had been allowed to cool about three minutes before stoppering, how might the results have been different? Why?

5. If the can had been stoppered about three minutes before heating was stopped, how might the results have been different? Why?

EXPERIMENT 21

Molar Volume of a Gas

MATERIALS AND EQUIPMENT

Solution: Hydrogen peroxide (3.0%, commercial preparation). **Solid:** manganese dioxide (powder MnO_2). **Special equipment:** Graduated cylinder, 50.0 ml; disposable syringe, 3.0 cc or 5.0 cc; stopper assembly with needle for attaching the syringe; battery jar or large beaker (1 or 2 L); large test tube; vacuum flask-Büchner funnel setup for MnO_2 disposal.

DISCUSSION

Often, gases must be handled and measured in the same experiment with solids and liquids. The amount of solid used or produced can be determined by measuring the mass of the material on a balance but it is difficult to measure the mass of a gas. It is much easier to measure gas volume and use this volume to calculate moles or mass. It is possible to do this because of Avogadro's law and the other gas laws. This experiment will illustrate the chemical significance of these important principles.

The term **molar volume** is used to describe the volume occupied by exactly one mole of any gas at a given temperature and pressure. Because a mole contains 6.022×10^{23} molecules (Avogadro's number), and Avogadro determined that equal volumes of different gases at the same temperature and pressure contain the same number of molecules (Avogadro's law) a mole of different gases will have exactly the same volume at the same temperature and pressure. The temperature and pressure at which molar volumes of gases are usually compared is standard temperature (0°C or 273 K) and standard pressure (1 atm). The molar volume of 22.4 L has been experimentally determined and verified over and over again for many gases in many laboratories. In this experiment, we will try to verify this value of 22.4 L for the molar volume of oxygen generated by the decomposition of hydrogen peroxide. The rate of this decomposition is greatly increased by the addition of MnO_2, a catalyst which remains unchanged by the reaction.

$$2\,H_2O_2 \xrightarrow{\text{MnO}_2} 2\,H_2O + O_2$$

If we think of molar volume as the liters per mole of gas at STP, then we need two measurements: the volume of oxygen at STP and the number of moles of oxygen that occupy that volume.

Using dimensional analysis and stoichiometry, the number of moles of oxygen gas generated can be calculated from the concentration and volume of H_2O_2. For example, if 2.0 ml of 5.0% hydrogen peroxide is decomposed in the reaction above, the number of moles of oxygen generated is calculated as follows:

$$\text{mol } O_2 = (2.0 \text{ mL } H_2O_2(aq))\left(\frac{5.0 \text{ g } H_2O_2}{100. \text{ mL } H_2O_2(aq)}\right)\left(\frac{1 \text{ mol } H_2O_2}{34.02 \text{ g } H_2O_2}\right)\left(\frac{1 \text{ mol } H_2O_2}{2 \text{ mol } H_2O_2}\right)$$

$$= 0.0015 \text{ mol } O_2$$

The oxygen will be collected by the downward displacement of water in an inverted graduated cylinder very much the way oxygen was collected in Experiment 3. The H_2O_2 solution is added to the generator with a syringe so there will be no loss of gas while putting the stopper in the tube after adding the solution. Figure 21.1 illustrates the setup for the generation and collection of oxygen in this experiment.

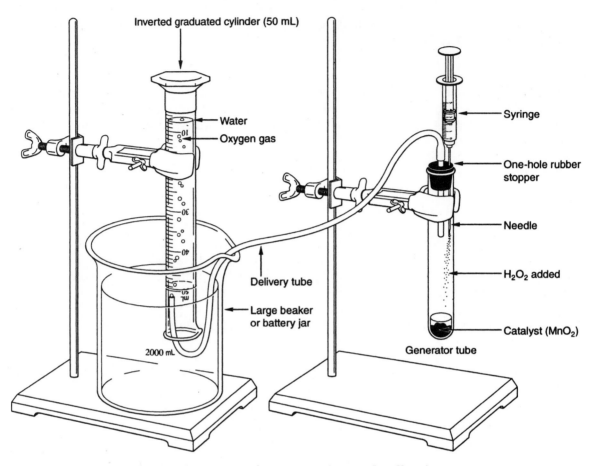

Figure 21.1   Apparatus for generating and collecting oxygen

The gas collected will contain some air because the oxygen generated will push air out of the generator and delivery tube and into the graduated cylinder. The mixture of air/oxygen/water vapor in the graduated cylinder will also include the air pushed out of the system by the addition of the hydrogen peroxide (2.0 mL in this example) so we must subtract the volume of peroxide added from the volume of gas collected to obtain the net volume of oxygen generated. Finally, before we can calculate the experimental molar volume the net volume of oxygen generated at laboratory conditions must be corrected to STP. This correction is done using the combined gas law:

$$\frac{V_1P_1}{T_1} = \frac{V_2P_2}{T_2} \qquad V_2 = \left(\frac{V_1P_1}{T_1}\right)\left(\frac{T_2}{P_2}\right)$$

Therefore, we must be able to measure V_1, T_1, and P_1 so we can solve for V_2.

V_1 is the volume of the gas collected after subtracting the volume of the H_2O_2 added. T_1 is the temperature in the laboratory changed to Kelvin. The pressure of the gas in the graduated cylinder is equal to the atmospheric pressure when the water levels inside and outside the graduate are the same. However, the gas collected also contains water vapor.

$$P_{O_2} + P_{H_2O} = P_{total} \quad \text{(atmospheric pressure)}$$

Therefore, to obtain the pressure of the oxygen collected (P_1) we need to subtract the vapor pressure of water from the atmospheric pressure. V_2 is the volume of oxygen at STP; it is calculated using the combined gas laws where P_2 is standard pressure (760 mm Hg) and T_2 is standard temperature (273 K).

Sample Calculation

When 2.0 mL of 5.0% H_2O_2 was reacted, 38.6 mL of gas was collected at 19.0°C. The vapor pressure of water at 19.0°C is 16.5 mm Hg (See Appendix 6). The barometric pressure was 743.5 mm Hg.

$V_1 = 38.6 \text{ mL} - 2.0 \text{ mL} = 36.6 \text{ mL}$

$T_1 = 19.0 + 273 = 292 \text{ K}$

$P_1 = 743.5 \text{ mm Hg} - 16.5 \text{ mm Hg} = 727.0 \text{ mm Hg}$

$T_2 = 273 \text{ K}$

$P_2 = 760 \text{ mm Hg}$

$$V_2 = \frac{V_1 P_1 T_2}{T_1 P_2} = \frac{(36.6 \text{ mL})(727.0 \text{ mm Hg})(273 \text{ K})}{(292 \text{ K})(760 \text{ mm Hg})} = 32.7 \text{ mL}$$

We now have all the information to determine the molar volume of O_2, 32.7 mL and 0.0015 mole O_2.

$$\frac{L}{mol} = \left(\frac{32.7 \text{ mL}}{0.0015 \text{ mol}}\right)\left(\frac{1 L}{1000 \text{ mL}}\right) = 22 \frac{L}{mol} \text{ (molar volume)}$$

If this experimental value is close to the theoretical value of 22.4 L/mol, we have verified Avogadro's law.

The percent error for our molar volume is determined as we have done in previous experiments:

$$\left(\frac{\text{Theoretical value} - \text{Experimental value}}{\text{Theoretical value}}\right)(100) = \left(\frac{22.4 \text{ L/mol} - 22 \text{ L/mol}}{22.4 \text{ L/mol}}\right)(100) = 1.8\%$$

PROCEDURE

Wear protective glasses.

1. Set up the apparatus as shown in Figure 21.1. Weigh about 2 g of MnO_2 and add it to the dry generator tube. It is not necessary to record this mass on the report form. Cover the MnO_2 with about 2 ml of distilled water and replace the stopper/syringe assembly. The needle should NOT be removed from the stopper at any time during this experiment. Make sure the stopper is firmly in the generator tube.

2. Fill the large beaker with water to about 1 inch from the top. The diameter of this container must accomodate a hand holding the inverted graduated cylinder so it must be fairly large. A 2 L beaker or battery jar works best. Fill the graduated cylinder to overflowing with water and cup your hand over the top being careful to keep all the water in the cylinder while you invert and submerge its top in the large beaker of water. Try to do this so there is no air at the "top" of the inverted cylinder when you are done. This may require several attempts and there will probably be one air bubble at the top of the graduate despite your best efforts. When successful, attach the inverted graduate to the ring stand as shown in Figure 21.1. Note that the clamp is attached near the top of the inverted cylinder. Do not put the delivery tube into the cylinder yet.

3. Determine the temperature of the gas to be collected by measuring the temperature in the laboratory. Then, read the mercury barometer to determine the atmospheric pressure in mm Hg. If a different type barometer is available, convert the units if necessary. Record these measurements on the report form.

4. Remove the syringe from the needle by twisting. It may be necessary to hold the needle with pliers when you twist the barrel. Pour about 10 mL of the peroxide solution into a small (50 mL) beaker and fill the syringe barrel to the 3.0 cc mark (1 cc = 1 mL). Do not replace the barrel on the needle yet. Accurately read and record the volume in the syringe. If you are unsure about reading the volume in the syringe, ask the instructor. Every significant digit is important here. Record the concentration of the peroxide solution on the report form. A solution which is 3.0% H_2O_2 (m/v) in water has 3.0 g of H_2O_2/100 ml of solution.

5. Carefully reattach the syringe to the needle. Hold the syringe by the barrel and be very careful to avoid moving the plunger prematurely. Put the gas delivery tube into the graduated cylinder. If a few air bubbles go into the cylinder at this point, make a note on your report form and proceed anyway. Inject the H_2O_2 into the generator by pushing firmly down on the syringe plunger. You will notice that air immediately bubbles up into the cylinder as you push the peroxide in. Then, as the peroxide and catalyst meet, the reaction begins immediately and proceeds slowly. Remove the entire generator tube from the ring stand and shake gently. It will take about two minutes to react all the H_2O_2. Be careful that the rubber stopper stays tightly in the top of the test tube and that the delivery tube stays in the graduated cylinder.

6. When the reaction is over, remove the delivery tube and carefully loosen the clamp holding the graduate to the ring stand. Move it up or down on the pole to equalize the levels of water inside and outside the graduate. Retighten the clamp to the ring stand. Read the volume of gas collected paying close attention to the meniscus. Although the gas in not pure oxygen, its volume is equal to the volume of oxygen generated plus the volume of air pushed out of the generator by the addition of the peroxide solution.

7. If you have any excess hydrogen peroxide solution, do not return it to the stock bottle. Pour it into the sink and flush generously with water.

8. Pour the contents of the generator tube into the vacuum flask—Büchner funnel setup for MnO_2 disposal. Return the stopper with needle and syringe to your instructor.

REPORT FOR EXPERIMENT 21

Molar Volume of a Gas

Measurements

1. Concentration of H_2O_2 _____ %

2. Volume of H_2O_2 added to the generator _____ mL

3. Barometric pressure _____ mm Hg

4. Water temperature _____ °C _____ K

5. Gas volume collected in the graduated cylinder _____ mL

Calculations: *Show the setup for every calculation.*

1. Oxygen volume generated by the reaction _____ mL

2. Moles of oxygen generated by the decomposition of H_2O_2 _____ mol

    (a) balanced equation for the reaction

    (b) stoichiometric setup using continuous calculation method

3. Pressure of dry gas in the graduated cylinder

    (a) vapor pressure of water _____ mm Hg

    (b) pressure of dry gas _____ mm Hg

4. Experimental volume of dry oxygen converted to STP _____ mL

5. Experimental molar volume _____ L/mol

6. Theoretical molar volume (L/mol) _____ L/mol

7. Percent error _____ %

QUESTIONS AND PROBLEMS

1. The curved surface of an aqueous solution is called _____. If the top of this surface were used to measure the volume of the gas rather than the bottom of the surface, the volume of the gas above the liquid would be _____

   (too large or too small)

2. Why was it necessary to subtract the volume of the H_2O_2 solution injected into the generator from the volume of gas collected in the graduated cylinder?

3. If a student did this experiment using 5.0 mL of 10.% hydrogen peroxide, how many moles of O_2 would be generated?

   _____ mol O_2

4. Why was it not necessary to measure the MnO_2 precisely when we have to be so careful about measuring the volume of H_2O_2 and the volume of oxgyen collected.

5. At 20.0°C, a student collects H_2 gas in a gas collecting tube. The barometric pressure is 755.2 mm Hg and the water levels inside and outside the tube are exactly equal.

   (a) What is the total gas pressure in the gas collecting tube?

   _____ mm Hg

   (b) What is the pressure of the water vapor in the gas collecting tube?

   _____ mm Hg

   (c) What is the pressure of the dry hydrogen in the gas collecting tube?

   _____ mm Hg

6. What would be the effect on the molar volume if hydrogen instead of oxygen gas had been collected during this experiment? Explain your answer.

7. After correction to standard conditions, the volume of a gas collected was 43.8 mL. If this volume represents 0.00184mol, what is the percent error for the experimental molar volume?

   _____ %

EXPERIMENT 22

Neutralization–Titration I

MATERIALS AND EQUIPMENT

Solid: potassium hydrogen phthalate, abbreviated KHP ($KHC_8H_4O_4$). **Liquids:** phenolphthalein indicator, unknown base solution (NaOH). One buret (25 mL or 50 mL) and buret clamp, buret brush. Wash bottle for distilled water.

DISCUSSION

The reaction of an acid and a base to form a salt and water is known as **neutralization.** In this experiment potassium hydrogen phthalate (abbreviated KHP) is used as the acid. Potassium hydrogen phthalate is an organic substance having the formula $HKC_8H_4O_4$, and like HCl, has only one acid hydrogen atom per molecule. Because of its complex formula, potassium hydrogen phthalate is commonly called KHP Despite its complex formula we see that the reaction of KHP with sodium hydroxide is similar to that of HCl. One mole of KHP reacts with one mole of NaOH.

$$HKC_8H_4O_4 + NaOH \longrightarrow NaKC_8H_4O_4 + H_2O$$

$$HCl + NaOH \longrightarrow NaCl + H_2O$$

Titration is the process of measuring the volume of one reagent required to react with a measured volume or mass of another reagent. In this experiment we will determine the molarity of a base (NaOH) solution from data obtained by titrating KHP with the base solution. The base solution is added from a buret to a flask containing a weighed sample of KHP dissolved in water. From the mass of KHP used we calculate the moles of KHP. Exactly the same number of moles of base is needed to neutralize this number of moles of KHP since one mole of NaOH reacts with one mole of KHP. We then calculate the molarity of the base solution from the titration volume and the number of moles of NaOH in that volume.

In the titration, the point of neutralization, called the **end-point,** is observed when an indicator, placed in the solution being titrated, changes color. The indicator selected is one that changes color when the stoichiometric quantity of base (according to the chemical equation) has been added to the acid. A solution of phenolphthalein, an organic acid, is used as the indicator in this experiment. Phenolphthalein is colorless in acid solution but changes to pink when the solution becomes slightly alkaline. When the number of moles of sodium hydroxide added is equal to the number of moles of KHP originally present, the reaction is complete. The next drop of sodium hydroxide added changes the indicator from colorless to pink.

Use the following relationships in your calculations:

1. According to the equation for the reaction,

    Moles of KHP reacted = Moles of NaOH reacted

2. $Moles = \dfrac{g \text{ of solute}}{molar \text{ mass of solute}}$

3. Molarity is an expression of concentration, the units of which are moles of solute per liter of solution:

$$\text{Molarity} = \frac{\text{moles}}{\text{liter}}$$

Thus, a 1.00 molar (1.00 M) solution contains 1.00 mole of solute in 1 liter of solution. A 0.100 M solution, then, contains 0.100 mole of solute in 1 liter of solution.

4. The number of moles of solute present in a known volume of solution of known concentration can be calculated by multiplying the volume of the solution (in liters) by the molarity of the solution:

$$\text{Moles} = (\text{liters})(\text{molarity}) = (\text{liters})\left(\frac{\text{moles}}{\text{liter}}\right)$$

PROCEDURE

Wear protective glasses.

Dispose of all solutions in the sink.

Make all weighings to the highest precision of the balance.

Obtain some solid KHP in a test tube or vial. Weigh two samples of KHP into 125 mL Erlenmeyer flasks, numbered for identification. (The flasks should be rinsed with distilled water, but need not be dry on the inside.) First weigh the flask, then add KHP to the flask by tapping the test tube or vial until 1.000 to 1.200 g has been added (see Figure 22.1). Determine the mass of the flask and the KHP. In a similar manner weigh another sample of KHP into the second flask. To each flask add approximately 30 mL of distilled water. If some KHP is sticking to the walls of the flask, rinse it down with water from a wash bottle. Warm the flasks slightly and swirl them until all the KHP is dissolved.

Figure 22.1   Method of adding KHP from a vial to a weighed Erlenmeyer flask

Obtain one buret and clean it. See "Use of the Buret," on the following page for instructions on cleaning and using the buret. Read and record all buret volumes to the nearest 0.01 mL.

Obtain about 250 mL of a base (NaOH) of unknown molarity in a clean, **dry** 250 mL Erlenmeyer flask as directed by your instructor. Record the number of this unknown.

1. Keep your base solution stoppered when not in use.

2. The 250 mL sample of base is intended to be used in both this experiment and Experiment 23. Be sure to label and save it.

Rinse the buret with two 5 to 10 mL portions of the base, running the second rinsing through the buret tip. Discard the rinsings in the sink. Fill the buret with the base, making sure that the tip is completely filled and contains no air bubbles. Adjust the level of the liquid in the buret so that the bottom of the meniscus is at exactly 0.00 mL. Record the initial buret reading (0.00 mL) in the space provided on the report form.

Add 3 drops of phenolphthalein solution to each 125 mL flask containing KHP and water. Place the first (Sample 1) on a piece of white paper under the buret extending the tip of the buret into the flask (see Figure 22.2).

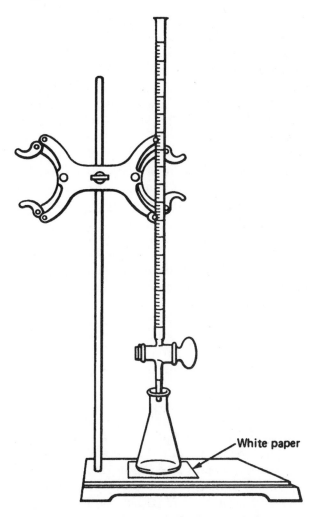

Figure 22.2   Setup with stopcock buret

Titrate the KHP by adding base until the end-point is reached. The titration is conducted by swirling the solution in the flask with the right hand (if you are right handed) while manipulating the stopcock with the left (Figure 22.3). As base is added you will observe a pink color caused by localized high base concentration. Toward the end-point the color flashes throughout the solution, remaining for a longer time. When this occurs, add the base drop by drop until the end-point is reached, as indicated by the first drop of base which causes a faint pink color to remain in the entire solution for at least 30 seconds. Read and record the final buret reading (see Figure 22.5). Refill the buret to the zero mark and repeat the titration with Sample 2. Then, calculate the molarity of the base in each sample. If these molarities differ by more than 0.004, titrate a third sample.

When you are finished with the titrations, empty and rinse the buret at least twice (including the tip) with tap water and once with distilled water. Return the vial with the unused KHP.

Use of the Buret

A buret is a volumetric instrument that is calibrated to deliver a measured volume of solution. The 50 mL buret is calibrated from 0 to 50 mL in 0.1 mL increments and is read to the nearest 0.01 mL. All volumes delivered from the buret should be between the calibration marks. (Do not estimate above the 0 mL mark or below the 50 mL mark.)

1. **Cleaning the Buret.**   The buret must be clean in order to deliver the calibrated volume. Drops of liquid clinging to the sides as the buret is drained are evidence of a dirty buret.

To clean the buret, first rinse it a couple of times with tap water, pouring the water from a beaker. Then scrub it with a detergent solution, using a long-handled buret brush. Rinse the buret several times with tap water and finally with distilled water. Check for cleanliness by draining the distilled water through the tip and observe whether droplets of water remain on the inner walls of the buret.

2. **Using the Buret.**   After draining the distilled water, rinse the buret with two 5 to 10 mL portions of the titrating solution to be used in it. This rinsing is done by holding the buret in a horizontal position and rolling the solution around to wet the entire inner surface. Allow the final rinsing to drain through the tip.

Fill the buret with the solution to slightly above the 0 mL mark and adjust it to 0.00 mL, or some other volume below this mark, by draining the solution through the tip. The buret tip must be completely filled to deliver the volume measured.

To deliver the solution from the buret, turn the stopcock with the forefinger and the thumb of your left hand (if you are right handed) to allow the solution to enter the flask. (See Figure 22.3). This procedure leaves your right hand free to swirl the solution in the flask during the titration. With a little practice you can control the flow so that increments as small as 1 drop of solution can be delivered.

3. **Reading the Buret.**   The smallest calibration mark of a 50 mL buret is 0.1 mL. However, the buret is read to the nearest 0.01 mL by estimating between the calibration marks. When reading the buret be sure your line of sight is level with the bottom of the meniscus in order to avoid parallax errors (see Figure 22.4). The exact bottom of the meniscus may be made more prominent and easier to read by allowing the meniscus to pick up the reflection from a heavy dark line on a piece of paper (see Figure 22.5).

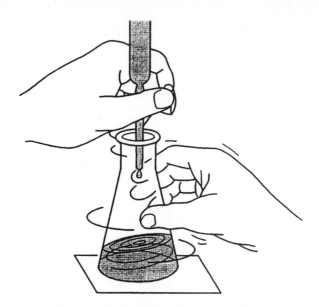

Figure 22.3   Titration technique

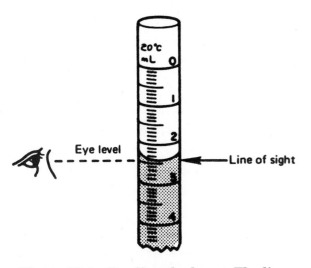

Figure 22.4   Reading the buret. The line of sight must be level with the bottom of the meniscus to avoid parallax.

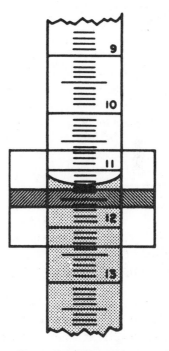

Figure 22.5   Reading the meniscus. A heavy dark line brought to within one division of the meniscus will make the meniscus more prominent and easier to read. The volume reading is 11.28 mL.

REPORT FOR EXPERIMENT 22

Neutralization – Titration I

Data Table

	Sample 1	Sample 2	Sample 3 (if needed)
Mass of flask and KHP			
Mass of empty flask			
Mass of KHP			
Final buret reading			
Initial buret reading			
Volume of base used			

CALCULATIONS: In the spaces below show calculation setups **for Sample 1 only.** Show answers for both samples in the boxes. Remember to use the proper number of significant figures in all calculations. (The number 0.005 has only one significant figure.)

	Sample 1	Sample 2	Sample 3 (if needed)
1. Moles of acid (KHP, Molar mass = 204.2)			
2. Moles of base used to neutralize (react with) the above number of moles of acid			
3. Molarity of base (NaOH)			

4. Average molarity of base _____

5. Unknown base number _____

QUESTIONS AND PROBLEMS

1. If you had added 50 mL of water to a sample of KHP instead of 30 mL, would the titration of that sample then have required more, less, or the same amount of base? Explain.

2. A student weighed out 1.106 g of KHP How many moles was that?

_____ mol

3. A titration required 18.38 mL of 0.1574 M NaOH solution. How many moles of NaOH were in this volume?

_____ mol

4. A student weighed a sample of KHP and found it weighed 1.276 g. Titration of this KHP required 19.84 mL of base (NaOH). Calculate the molarity of the base.

_____ M

5. Forgetful Freddy weighed his KHP sample, but forgot to bring his report sheet along, so he recorded the mass of KHP on a paper towel. During his titration, which required 18.46 mL of base, he spilled some base on his hands. He remembered to wash his hands, but forgot about the data on the towel, and used it to dry his hands. When he went to calculate the molarity of his base, Freddy discovered that he didn't have the mass of his KHP. His kindhearted instructor told Freddy that his base was 0.2987 M. Calculate the mass of Freddy's KHP sample.

_____ g

6. What mass of solid NaOH would be needed to make 645 mL of Freddy's NaOH solution?

_____ g

EXPERIMENT 23

Neutralization – Titration II

MATERIALS AND EQUIPMENT

Solutions: Acid of unknown molarity, standard base solution (NaOH), vinegar, phenolphthalein indicator. Suction bulb, buret, buret brush, buret clamp, 10 mL volumetric pipet. Wash bottle for distilled water.

DISCUSSION

This experiment may follow Experiment 22 or it may be completed independently of Experiment 22. In either case the discussion section of Experiment 22 supplements the following discussion.

The reaction of an acid and a base to form water and a salt is known as **neutralization.** Hydrochloric acid and sodium hydroxide, for example, react to form sodium chloride and water.

$$HCl(aq) + NaOH(aq) \longrightarrow H_2O(l) + NaCl(aq)$$

The ionic reaction in neutralizations of this type is that of hydrogen (or hydronium) ion reacting with hydroxide ion to form water.

$$H^+(aq) + OH^-(aq) \longrightarrow H_2O(l) \quad \text{or} \quad H_3O^+(aq) + OH^-(aq) \longrightarrow 2\,H_2O(l)$$

A monoprotic acid—i.e., an acid having one ionizable hydrogen atom per molecule—reacts with sodium hydroxide (or any other monohydroxy base) on a 1:1 mole basis. This fact is often utilized in determining the concentrations of solutions of acids by titration.

Titration is the process of measuring the volume of one reagent to react with a measured volume or mass of another reagent. In this experiment an acid solution of unknown concentration is titrated with a base solution of known concentration, Phenolphthalein is used as an indicator. This substance is colorless in acid solution, but changes to pink when the solution becomes slightly basic or alkaline. The change of color, caused by a single drop of the base solution in excess over that required to neutralize the acid, marks the **end-point** of the titration.

Molarity (M) is the concentration of a solution expressed in terms of moles of solute per liter of solution.

$$\text{Molarity} = \frac{\text{moles}}{\text{liter}}$$

Thus a solution containing 1.00 mole of solute in 1.00 liter of solution is 1.00 molar (1.00 M). If only 0.155 mole is present in 1.00 liter of solution, it is 0.155 M, etc. To determine the molarity of any quantity it is only necessary to divide the total number of moles of solute present in the solution by the volume (in liters).

To determine the number of moles of solute present in a known volume of solution, multiply the volume in liters by the molarity.

$$\text{Moles} = (\text{liters})(\text{molarity}) = (\text{liters})\left(\frac{\text{moles}}{\text{liter}}\right)$$

For titrations involving monoprotic acids and monohydroxy bases (one hydroxide ion per formula unit), the number of moles of acid is identical to the number of moles of base required to neutralize the acid. In this experiment we measure the volume of base of known molarity required to neutralize a measured volume of acid of unknown molarity. The molarity of the acid can then be calculated.

$$\text{Moles base} = (\text{liters})(\text{molarity}) = (\text{liters base})\left(\frac{\text{moles base}}{\text{liters}}\right)$$

$$\text{Moles acid} = (\text{moles base})\left(\frac{1 \text{ mole acid}}{1 \text{ mole base}}\right)$$

$$\text{Molarity of acid} = \frac{\text{moles acid}}{\text{liters acid}}$$

In order to determine the molarity of an acid solution, it is not actually necessary to know what the acid is—only whether it is monoprotic, diprotic, or triprotic. The calculations in this experiment are based on the assumption that the acid in the unknown is monoprotic.

If the molarity and the formula of the solute are known, the concentration in grams of solute per liter of the solution may be calculated by multiplying by the molar mass.

$$(\text{Molarity})(\text{molar mass}) = \left(\frac{\text{moles}}{\text{liter}}\right)\left(\frac{\text{grams}}{\text{mole}}\right) = \frac{\text{grams}}{\text{liter}}$$

In determining the acid content of commercial vinegar, it is customary to treat the vinegar as a dilute solution of acetic acid, $HC_2H_3O_2$. The acetic acid concentration of the vinegar may be calculated as grams of acetic acid per liter or as percent acid by mass. If the acetic acid content is to be expressed on a mass percent basis, the density of the vinegar must also be known.

PROCEDURE

Wear protective glasses.

Do not pipet by mouth.

 Dispose of all solutions in the sink. Flush with water.

A. Molarity of an Unknown Acid

Obtain a sample of acid of unknown molarity in a clean, dry 125 mL Erlenmeyer flask as directed by your instructor.

With a volumetric pipet, transfer a 10.00 mL sample of the acid to a clean, but not necessarily dry, Erlenmeyer flask. See "Use of the Pipet," on the following page, for instructions on cleaning and using the pipet. Pipet a duplicate 10.00 mL sample into a second flask. (If pipets are not available, a buret which has been carefully cleaned and rinsed may be used to measure the acid samples.)

You will need about 150 mL of base of known molarity (standard solution). Your instructor will give you the exact molarity of the base solution that you used in Experiment 22 or you may be given another sodium hydroxide solution of known molarity. Record the exact molarity of this solution. Keep the flask containing the base stoppered when not in use.

Clean and set up a buret. See "Use of the Buret," in Experiment 22, for instructions on cleaning and using the buret.

Rinse the buret with two 5 to 10 mL portions of the base, running the second rinsing through the buret tip. Discard the rinsings in the sink. Fill the buret with the base, making sure that the tip is completely filled and contains no air bubbles. Adjust the level of the liquid in the buret so that the bottom of the meniscus is near or exactly at 0.00 mL. Record the initial buret reading in the space provided on the report form.

Add three drops of phenolphthalein solution and about 25 mL of distilled water to the flask containing the 10.00 mL of acid. Place this flask on a piece of white paper under the buret and lower the buret tip into the flask (see Figure 22.2).

Titrate the acid by adding base until the end-point is reached. During the titration swirl the solution in the flask with the right hand (if you are right handed) while manipulating the stopcock with the left. As the base is added you will observe a pink color caused by localized high base concentration. Near the end-point this color flashes throughout the solution, remaining for increasingly longer periods of time. When this occurs, add the base drop by drop until the end-point is reached, as indicated by the first drop of base which causes the entire solution to retain a faint pink color for at least 30 seconds. Record the final buret reading.

Refill the buret with base and adjust the volume to near the zero mark. Titrate the duplicate sample of acid. If the volumes of base used differ by more than 0.20 mL, titrate a third sample. In the calculations, assume that the unknown acid reacts like KHP or HCl (one mole of acid reacts with one mole of base).

B. Acetic Acid Content of Vinegar

Obtain about 40 mL of vinegar in a clean, dry 50 mL beaker. Record the sample number, if any, of this vinegar.

Titrate duplicate 10.00 mL samples of vinegar using exactly the same procedure outlined in Part A. Remember to rinse the pipet with vinegar before pipeting the vinegar samples.

When you are finished with the titrations, empty the buret and rinse it and the pipet at least twice with tap water and once with distilled water.

Use of the Pipet

A volumetric (transfer) pipet (Figure 23.1) is calibrated to deliver a specified volume of liquid to a precision of about ±0.02 mL in a 10 mL pipet. To achieve this precision, the pipet must be clean and used in a specified manner.

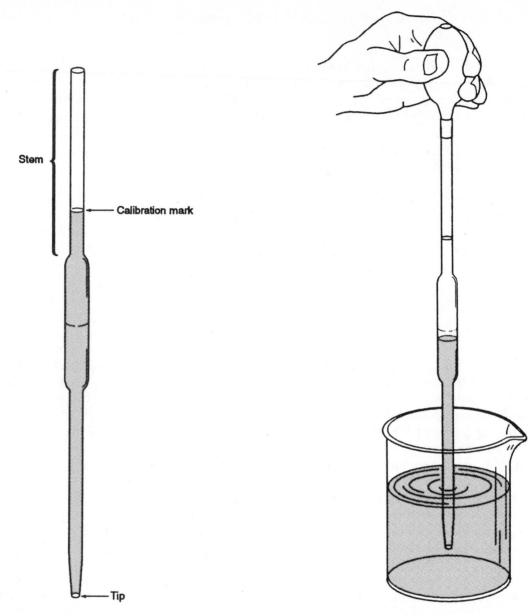

Figure 23.1   A volumetric (transfer) pipet

Figure 23.2   Liquid is drawn into the pipet with a rubber suction bulb. Keep the tip of the pipet below the liquid level during suction.

Liquids are drawn into a pipet by means of a rubber suction bulb (Figure 23.2) or by a rubber tube connected to a water aspirator pump. Suction by mouth has also been used to draw liquids into a pipet, but this is a dangerous practice and is not recommended.

1. **Cleaning the Pipet.**   Use a rubber suction bulb to draw up enough detergent solution to fill about two-thirds of the body or bulb of the pipet. Retain this solution in the pipet by pressing the forefinger tightly against the top of the pipet stem (Figure 23.3). turn the pipet to a nearly horizontal position and gently shake and rotate it until the entire inside surface is wetted. Allow the pipet to drain and rinse it at least three times with tap water and once with distilled water.

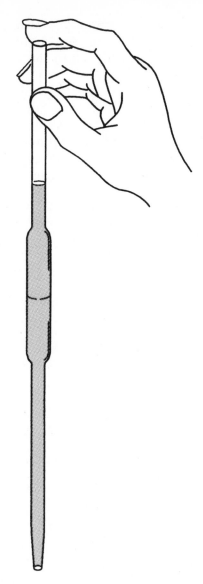

Figure 23.3   Liquid is retained in the pipet by applying pressure with the forefinger to the top of the stem.

Figure 23.4   The pipet is calibrated to deliver the specified volume, leaving a small amount of liquid in the tip.

2. **Using the Pipet.**   Unless the pipet is known to be clean and absolutely dry on the inside, it must be rinsed twice with small portions of the liquid that is to be pipeted. This is done as in the washing procedure described above. These rinses are discarded in order to avoid contamination of the liquid being pipeted. A **pipet** does not need to be rinsed between successive pipettings of the same solution.

To transfer a measured volume of a liquid, collapse a suction bulb by squeezing and place it tightly against the top of a pipet. (Do not try to push the bulb on to the pipet.) Draw the liquid into the pipet until it is filled to about 5 cm above the calibration mark by allowing the bulb to slowly expand. Be careful—do not allow the liquid to get into the bulb. Remove the bulb and quickly place your forefinger over the top of the pipet stem. The liquid will be retained in the pipet if the finger is pressed tightly against the top of the stem. Keeping the pipet in a vertical position, decrease the finger pressure very slightly, and allow the liquid level to drop slowly toward the calibration mark. When the liquid level has almost reached the calibration mark, again increase the finger pressure and stop the liquid when the bottom of the meniscus is exactly on the calibration mark. Touch the tip to the wall of the flask to remove the adhering drop of liquid.

Move the pipet to the flask which is to receive the sample and allow the liquid to drain while holding the pipet in a vertical position. About 10 seconds after the liquid has stopped running from the pipet, touch the tip to the inner wall of the sample flask to remove the drop of liquid adhering to the tip. A small amount of liquid will remain in the tip (Figure 23.4). Do not blow or shake this liquid into the sample; the pipet is calibrated to deliver the volume specified without this small residual.

If you have never used a volumetric pipet, it is advisable to practice by pipetting some samples of distilled water until you have mastered the technique.

NAME ____________________

SECTION ______ DATE ______

REPORT FOR EXPERIMENT 23

INSTRUCTOR____________________

Neutralization-Titration II

A. Molarity of an Unknown Acid

Data Table

	Sample 1		**Sample 2**		**Sample 3 (if needed)**	
	Acid*	Base	Acid*	Base	Acid*	Base
Final buret reading						
Initial buret reading						
Volume used						

*If a pipet is used to measure the volume of acid, record only in the space for volume used.

Molarity of base (NaOH) ________________

CALCULATIONS: In the spaces below, show calculation setups for Sample 1 only. Show answers for both samples in the boxes.

	Sample 1	**Sample 2**	**Sample 3 (if needed)**
1. Moles of base (NaOH) (if needed)			
2. Moles of acid used to neutralize (react with) the above number of moles of base			
3. Molarity of acid			

4. Average molarity of acid ____________________

5. Unknown acid number ____________________

B.   **Acetic Acid Content of Vinegar**

Data Table

	Sample 1		Sample 2		Sample 3 (if needed)	
	Vinegar*	Base	Vinegar*	Base	Vinegar*	Base
Final buret reading						
Initial buret reading						
Volume used						

*If a pipet is used to measure the volume of vinegar, record only in the space for volume used.

Molarity of base (NaOH) _____   Vinegar number _____

CALCULATIONS: In the spaces below, show calculation setups for Sample 1 only. Show answers for both samples in the boxes.

	Sample 1	Sample 2	Sample 3 (if needed)
1. Moles of base (NaOH)			
2. Moles of acid ($HC_2H_3O_2$) used to neutralize (react with) the above number of moles of base			
3. Molarity of acetic acid in the vinegar			

4. Average molarity of acetic acid in the vinegar   _____

5. Grams of acetic acid per liter (from average molarity)   _____

6. Mass percent acetic acid in vinegar sample
    (density of vinegar = 1.005 g/mL)   _____

EXPERIMENT 24

Chemical Equilibrium – Reversible Reactions

MATERIALS AND EQUIPMENT

Solid: ammonium chloride (NH_4Cl). **Solutions:** saturated ammonium chloride, concentrated (15 M) ammonium hydroxide (NH_4OH), 0.1 M cobalt(II) chloride ($CoCl_2$), 0.1 M iron(III) chloride ($FeCl_3$), concentrated (12 M) hydrochloric acid (HCl), 0.1 M copper(II) sulfate ($CuSO_4$), 6 M ammonium hydroxide (NH_4OH), phenolphthalein, 0.1 M potassium thiocyanate (KSCN), 0.1 M silver nitrate ($AgNO_3$), saturated sodium chloride (NaCl), and dilute (3 M) sulfuric acid (H_2SO_4).

DISCUSSION

In many chemical reactions the reactants are not totally converted to the products because of a reverse reaction; that is, because the products react to form the original reactants. Such reactions are said to be reversible and are indicated by a double arrow ($\rightleftharpoons$) in the equation. The reaction proceeding to the right is called the **forward reaction;** that to the left, the **reverse reaction.** Both reactions occur simultaneously.

Every chemical reaction proceeds at a certain rate or speed. The rate of a reaction is variable and depends on the concentrations of the reactants and the conditions under which the reaction is conducted. When the rate of the forward reaction is equal to the rate of the reverse reaction, a condition of **chemical equilibrium** exists. At equilibrium the products react at the same rate as they are produced. Thus the concentrations of substances in equilibrium do not change, but both reactions, forward and reverse, are still occurring.

The principle of Le Chatelier relates to systems in equilibrium and states that when the conditions of a system in equilibrium are changed the system reacts to counteract the change and reestablish equilibrium. In this experiment we will observe the effect of changing the concentration of one or more substances in a chemical equilibrium. Consider the hypothetical equilibrium system

$$A + B \rightleftharpoons C + D$$

When the concentration of any one of the species in this equilibrium is changed, the equilibrium is disturbed. Changes in the concentrations of all the other substances will occur to establish a new position of equilibrium. For example, when the concentration of B is increased, the rate of the forward reaction increases, the concentration of A decreases, and the concentrations of C and D increase. After a period of time the two rates will become equal and the system will again be in equilibrium. The following statements indicate how the equilibrium will shift when the concentrations of A, B, C, and D are changed.

An increase in the concentration of A or B causes the equilibrium to shift to the right.

An increase in the concentration of C or D causes the equilibrium to shift to the left.

A decrease in the concentration of A or B causes the equilibrium to shift to the left.

A decrease in the concentration of C or D causes the equilibrium to shift to the right.

Evidence of a shift in equilibrium by a change in concentration can easily be observed if one of the substances involved in the equilibrium is colored. The appearance of a precipitate or the change in color of an indicator can sometimes be used to detect a shift in equilibrium.

Net ionic equations for the equilibrium systems to be studied are given below. These equations will be useful for answering the questions in the report form.

A. Saturated Sodium Chloride Solution

$$NaCl(s) \overset{H_2O}{\rightleftharpoons} Na^+(aq) + Cl^-(aq)$$

B. Saturated Ammonium Chloride Solution

$$NH_4Cl(s) \overset{H_2O}{\rightleftharpoons} NH_4^+(aq) + Cl^-(aq)$$

C. Iron(III) Chloride plus Potassium Thiocyanate

$$Fe^{3+}(aq) + SCN^-(aq) \rightleftharpoons Fe(SCN)^{2+}(aq)$$

Pale yellow   Colorless   Red

D. Copper(II) Sulfate Solution with Ammonia

$$Cu(H_2O)_4^{2+}(aq) + 4\,NH_3(aq) \rightleftharpoons Cu(OH)_2(s) \rightleftharpoons [Cu(NH_3)_4]^{2+}(aq) + 4\,H_2O$$

light blue clear   J   blue cloudy   K   deep blue/purple cloudy

E. Cobalt(II) Chloride Solution

The equilibrium involves the following ions in solutions:

$$Co(H_2O)_6^{2+}(aq) + 4\,Cl^-(aq) \rightleftharpoons CoCl_4^{2-}(aq) + 6\,H_2O(l)$$

Pink   Blue

F. Ammonia Solution

$$NH_3(aq) + H_2O(l) \rightleftharpoons NH_4^+(aq) + OH^-(aq)$$

PROCEDURE

Wear protective glasses.

NOTE: Record observed evidence of equilibrium shifts as each experiment is done.

A. Saturated Sodium Chloride Solution

Add a few drops of conc. hydrochloric acid to 2 to 3 mL of saturated sodium chloride solution in a test tube, and note the results.

B. Saturated Ammonium Chloride Solution

Repeat Part A, using saturated ammonium chloride solution instead of sodium chloride solution.

 Dispose of the solutions in A and B in the sink and flush with water.

C. Iron(III) Chloride plus Potassium Thiocyanate

Prepare a stock solution to be tested by adding 2 mL each of 0.1 M iron(III) chloride and 0.1 M potassium thiocyanate solutions to 100 mL of distilled water and mix. Pour about 5 mL of this stock solution into each of four test tubes.

1. Use the first tube as a control for color comparison.

2. Add about 1 mL of 0.1 M iron(III) chloride solution to the second tube and observe the color change.

3. Add about 1 mL of 0.1 M potassium thiocyanate solution to the third tube and observe the color change.

4. Add 0.1 M silver nitrate solution dropwise (less than 1 mL) to the fourth tube until almost all the color is discharged. The white precipitate formed consists of both $AgCl$ and $AgSCN$. Pour about half the contents (including the precipitate) into another tube. Add 0.1 M potassium thiocyanate solution dropwise (1 to 2 mL) to one tube and 0.1 M iron(III) chloride solution (1 to 2 mL) to the other. Observe the results.

 Dispose of the contents in tubes C.1–3 and the unused stock solutions in the sink and flush with water. Dispose of the contents of both C.4 tubes in the "heavy metals" waste container.

D. Copper (II) Sulfate Solution with Ammonia

Pour 2 mL of 0.1 M copper (II) sulfate into each of two test tubes. Add 6 M $NH_3(aq)$ (NH_4OH) dropwise (shake well after each drop is added) to one of the copper(II) sulfate tubes. When there is a definite color or appearance change, note the change on the report form. Use the second test tube for comparison. Continue to add the $NH_3(aq)$ until there is another color or appearance change. Note the changes on the report form.

Now, add 3 M H_2SO_4 dropwise to the solution until the original color is restored. Again, use the second tube for comparison.

 Dispose of the contents of both test tubes in the "heavy metals" waste container provided.

E. Cobalt(II) Chloride Solution

Place about 2 mL (no more) of 0.1 M cobalt(II) chloride solution into each of three test tubes.

   1. To one tube add about 3 mL of conc. hydrochloric acid dropwise and note the result. Now add water dropwise to the solution until the original color (reverse reaction) is evident.

   2. To the second tube add about 1.5 g of solid ammonium chloride and shake to make a saturated salt solution. Compare the color with the solution in the third tube (control). Place the second and third tubes (unstoppered) in a beaker of boiling water, shake occasionally, and note the results. Cool both tubes under tap water until the original color (reverse reaction) is evident.

 Dispose of these solutions in the "heavy metals" waste container.

F. Ammonia Solution

Prepare an ammonia stock solution by adding 10 drops of 6 M ammonium hydroxide and 3 drops of phenolphthalein to 100 mL of tap water and mix. Pour about 5 mL of this stock solution into each of two test tubes.

   1. Dissolve a very small amount of solid ammonium chloride in the stock solution in the first tube and observe the result.

   2. Add a few drops of dil. (6 M) hydrochloric acid to the stock solution in the second tube. Mix and observe the result.

 Dispose of these solutions and the rest of the ammonia stock solution in the sink and flush with water.

NAME _____

SECTION _____ DATE _____

INSTRUCTOR _____

REPORT FOR EXPERIMENT 24

Chemical Equilibrium – Reversible Reactions

Refer to equilibrium equations in the discussion when answering these questions.

A.   Saturated Sodium Chloride

1. What is the evidence for a shift in equilibrium?

2. Which ion caused the equilibrium to shift?   _____

3. In which direction did the equilibrium shift?   _____

4. If solid sodium hydroxide were added to neutralize the hydrochloric acid, would this reverse the reaction and cause the precipitated sodium chloride to redissolve? Explain.

B.   Saturated Ammonium Chloride

1. What is the evidence for a shift in equilibrium?

2. In which direction did the equilibrium shift?   _____

3. Which ion caused the equilibrium to shift?   _____

C.   Iron(III) Chloride plus Potassium Thiocyanate

1. What is the evidence for a shift in equilibrium when iron(III) chloride is added to the stock solution?

2. What is the evidence for a shift in equilibrium when potassium thiocyanate is added to the stock solution?

3. (a) What is the evidence for a shift in equilibrium when silver nitrate is added to the stock solution? (The formation of a precipitate is not the evidence since the precipitate is not one of the substances in the equilibrium.)

(b) The change in concentration of which ion in the equilibrium caused this equilibrium shift?

_____

(c) Write a net ionic equation to illustrate how this concentration change occurred.

(d) When the mixture in C.4 was divided and further tested, what evidence showed that the mixture still contained Fe^{3+} ions in solution?

D. Copper(II) Sulfate Solution with Ammonia

1. What was the evidence for the first shift in equilibrium when the $NH_3(aq)$ was added dropwise to the Cu^{2+} solution?

2. (a) Explain how adding more $NH_3(aq)$ caused the equilibria to shift again.

(b) What did you observe in the Cu^{2+} system to indicate that the shift had occured?

3. (a) Explain how 3 M sulfuric acid caused the equilibria to shift back again?

(b) What did you observe to indicate that the reaction shifted to the left?

E.   Cobalt(II) Chloride Solution

1. What was the evidence for a shift in equilibrium when conc. hydrochloric acid was added to the cobalt chloride solution?

2. (a) Write the equilibrium equation for this system.

    (b) State whether the concentration of each of the following substances was increased, decreased, or unaffected when the conc. hydrochloric acid was added to cobalt chloride solution.

    $Co(H_2O)_6^{2+}$ _____, Cl^- _____, $CoCl_4^{2-}$ _____

3. (a) What did you observe when ammonium chloride was added to cobalt chloride solution?

    (b) What did you observe when this mixture was heated?

    (c) Explain why heating the mixture caused the equilibrium to shift.

    (d) What did you observe when the mixture was cooled?

    (e) Explain why cooling the mixture caused the equilibrium to shift.

F. Ammonia Solution

1. What is the evidence for a shift in equilibrium when ammonium chloride was added to the stock solution?

2. Explain, in terms of the equilibrium, the results observed when hydrochloric acid was added to the stock solution.

3. State whether the concentration of each of the following was increased, decreased, or was unaffected when dilute hydrochloric acid was added to the ammonia stock solution:

    NH_3 _____, NH_4^+ _____, OH^- _____,   Pink color _____

4. (a) In which direction would the equilibrium shift if sodium hydroxide were added to the ammonia stock solution?

    (b) Would the sodium hydroxide tend to decrease the color intensity? Explain.

5. Would boiling the ammonia solution have any effect on the equilibrium? Explain.

EXPERIMENT 25

Heat of Reaction

MATERIALS AND EQUIPMENT

Solid: ammonium chloride (NH_4Cl); **Solutions:** 3 M hydrochloric acid (HCl), 3 M nitric acid (HNO_3), 1.25 M sodium hydroxide (NaOH), and concentrated (18 M) sulfuric acid (H_2SO_4). Styrofoam cups; thermometer.

DISCUSSION

All chemical reactions involve a heat effect, called the **heat of reaction.** If the reaction liberates heat, the heat effect is an increase in temperature and the reaction is **exothermic.** If the reaction consumes heat, the heat effect is a decrease in temperature and it is called an **endothermic** reaction. The heat effect of any given reaction is the result of the formation and breaking of chemical bonds during the reaction. Energy is consumed when bonds are broken; energy is liberated in the formation of bonds. The **heat of reaction** is often the combination of several effects. Units used to quantify heats of reaction are **joules** or **kilojoules** which are determined indirectly using **calorimetry** from several direct measurements, including temperature change. Review Experiment 5 for a review of calorimetry techniques.

This experiment will investigate the heat of reaction for three types of reactions, all of which are conducted in an aqueous solution with a styrofoam cup serving as a calorimeter:

A. The hydration of a liquid

B. The dissolving of a solid, and

C. The neutralization of an acid with a base.

The quantity of heat for each reaction will not be measured directly in any of the experiments. Instead temperature changes will be observed as indicators of heat effects. From the temperature changes observed, and the specific heat of the water, the heat liberated or consumed can be calculated.

A. Hydration of Concentrated Sulfuric Acid

A strong heat effect is observed when concentrated (18 M) sulfuric acid is added to water. This reaction is so exothermic that it can lead to dangerous spattering. The general rule "add the acid to the water" will help avoid the danger.

The reaction involved is the hydration of sulfuric acid molecules to form $H_2SO_4 \cdot H_2O$ and $H_2SO_4 \cdot 2H_2O$. The formation of the bonds between the acid and the water is the primary cause of the heat liberated, known as **heat of hydration,** and the resulting temperature increase. The usual commercial concentrated sulfuric acid contains less than 5 percent water, so some of the acid molecules are already hydrated.

B. Dissolving Ammonium Chloride

When a salt dissolves, there are two major competing heat effects. The energy required to break the ionic bonds of the crystal lattice is called the **lattice energy.** As the ions are freed from the crystal lattice they become hydrated, liberating hydration energy as described for the hydration of sulfuric acid. If the lattice energy is greater than the hydration energy, the reaction will be endothermic. If the hydration energy is greater than the lattice energy, the reaction will be exothermic. In this experiment, ammonium chloride will be dissolved in water and the temperature change will be measured to determine which is greater, the hydration energy or the lattice energy involved in the solution process.

C. Heat of Neutralization

When hydrochloric acid is added to sodium hydroxide solution, heat is evolved. Since the reaction of an acid with a base is called neutralization, this energy is called the **heat of neutralization.** The molecular equation for this reaction is:

$$HCl(aq) + NaOH(aq) \longrightarrow H_2O(l) + NaCl(aq) + heat$$

Since all the reactants and products are soluble, and all of them except water are strong electrolytes, the total ionic equation for this reaction is

$$H^+(aq) + Cl^-(aq) + Na^+(aq) + OH^-(aq) \longrightarrow H_2O(l) + Na^+(aq) + Cl^-(aq) + heat$$

Therefore, the net ionic equation for this neutralization reaction is:

$$\underset{\text{1 mole}}{H^+(aq)} + \underset{\text{1 mole}}{OH^-(aq)} \longrightarrow \underset{\text{1 mole}}{H_2O(l)} + heat$$

According to this net ionic equation, a neutralization reaction between a strong acid and a strong base is the reaction of the hydrogen ions (H^+) and the hydroxide ions (OH^-) to form water. Since this reaction is common to all strong acids and bases, the amount of heat liberated per mole of water is known as the heat of neutralization, and has been determined to be 55.9 kJ/mole. Thus, it should be possible to use different strong acids with a fixed amount of base and obtain approximately the same quantity of heat for each acid neutralized. Therefore, we can write the neutralization reaction as

$$H^+(aq) + OH^-(aq) \longrightarrow H_2O(l) + 55.9\,kJ$$

In this experiment, we will use different strong acids with a fixed amount of base and measure the temperature increase which results for each reaction system. Each neutralization will involve the same total volume of solution. Thus, if the same quantity of heat actually is produced, the temperature increase should be the same for each neutralization. This assumes a constant specific heat (4.184 J/g°C) for the various aqueous solutions used, which is not precisely true, but is satisfactory for the purpose of this experiment. To minimize experimental errors in measuring temperature, two trials with each acid will be done and dilute acids and bases are used to minimize heats of hydration.

In all the experimental trials the moles of hydroxide ion will be the limiting reactant. Thus, there will always be more moles of hydrogen ion available than of hydroxide ions. The excess moles of hydrogen ion will remain unreacted.

To determine the accuracy of the heat effects observed for these neutralization reactions, we will calculate the kJ/mole of water formed using the specific heat of water, the average temperature change for all the trials measured, and the mass of the solution in the calorimetry equation:

$$q = (m)(sp.ht)(\Delta t)$$

where q = heat liberated
m = mass of solution
$sp.\ ht.$ = specific heat of water
Δt = temperature change

Sample calculation: A 40.0 mL sample of 3.0 M HCl is added to a calorimeter cup and the temperature is measured at 20.0°C. 35.0 mL of 1.25 M NaOH is added and stirred with a thermometer and the maximum temperature obtained is 27.4°C. Assume the density of the solution is 1.04 g/mL.

a. Calculate the heat produced by the reaction.

$$q = (75.0\,\text{mL})\left(\frac{1.04\,\text{g}}{1\,\text{mL}}\right)\left(\frac{4.184\,\text{J}}{\text{g°C}}\right)(7.4°C) = 2.4 \times 10^3\,\text{J}$$

b. How many moles of water were produced?

$$\text{mol}\ H_2O = (35.0\,\text{mL NaOH})\left(\frac{1.25\,\text{mol NaOH}}{1000\,\text{mL NaOH}}\right)\left(\frac{1\,\text{mol}\ H_2O}{1\,\text{mol NaOH}}\right) = 0.0438\,\text{mol}\ H_2O$$

c. What is the amount of heat (in kJ/mol) that would be liberated if 1 mole of H_2O had been formed?

$$\text{kJ/mol} = \left(\frac{2.4 \times 10^3\,\text{J}}{0.0438\,\text{mol}}\right)\left(\frac{1\,\text{kJ}}{1000\,\text{J}}\right) = 55\,\text{kJ/mol}\ H_2O\ \text{formed}$$

d. What is the percent error for this experimental value (vs. the accepted value of 55.9 kJ/mol)?

$$\left(\frac{\text{Accepted value} - \text{Experimental value}}{\text{Accepted value}}\right)(100) = \left(\frac{55.9\,\text{kJ/mol} - 55\,\text{kJ/mol}}{55.9\,\text{kJ/mol}}\right)(100) = 1.6\%$$

PROCEDURE

⚠ Wear protective glasses

WASTE
DISPOSE OF
PROPERLY
  Dispose of all solutions in the sink and flush with water.

Record all temperatures to the nearest 0.1°C.

A. Hydration of Concentrated Sulfuric Acid

Concentrated sulfuric acid is an extremely corrosive oxidizing acid and a strong dehydrating agent. Wash immediately if you spill any on your skin or clothing. Rinse off any acid that runs down the side of the bottle with a wash bottle and rinse the graduated cylinder used.

Pour 45 mL of tap water into a styrofoam cup. Read and record the temperature of the water. Measure 5.0 mL of concentrated sulfuric acid and pour it into the water in the styrofoam cup. Stir with the thermometer and record the maximum temperature reached. Pour this dilute sulfuric acid solution into a 125 mL Erlenmeyer flask and save it for Part C.

B. Dissolving of Ammonium Chloride

Rinse the styrofoam cup and pour 20 mL of tap water into it. Read and record the temperature of the water. Weigh approximately 3 g of ammonium chloride crystals and pour them into the styrofoam cup. Stir with the thermometer and record the minimum temperature reached.

 Pour the ammonium chloride solution down the sink and flush generously with water. Rinse the cup and reuse for Part C.

C. Neutralization Reactions

You will need five stock solutions which should be placed in small, labeled Erlenmeyer flasks as indicated below:

1. 140 mL of 1.25 M NaOH

2. 100 mL of 3 M HCl

3. 30 mL of 3 M HNO_3

4. H_2SO_4 solution prepared in Part A. (1.8 M)

5. 180 mL of distilled water at room temperature

After you have the stock solutions assembled, you will carry out five neutralization reactions in the Styrofoam cup and measure the maximum temperature reached for each reaction. For each reaction:

1. Add the measured volume of the specified stock acid and water to the styrofoam cup and stir.

2. Record the temperature of the acid solution.

3. Add the measured volume of the base (NaOH), stir with the thermometer, and record the maximum temperature reached.

4. Do the second trial with the same acid now.

5. Empty the styrofoam cup, rinse the cup, and graduated cylinders with water.

6. Repeat these steps for the remaining neutralization reactions using the quantities shown:

Neutralization #1	#2	#3	#4	#5
20.0 mL H_2O	20.0 mL H_2O	20.0 mL H_2O	10.0 mL H_2O	10.0 mL H_2O
10.0 mL HCl	10.0 mL HNO_3	10.0 mL H_2SO_4	20.0 mL HCl	10.0 mL HCl
10.0 mL NaOH	10.0 mL NaOH	10.0 mL NaOH	10.0 mL NaOH	20.0 mL NaOH

NAME ____________________

SECTION ________ DATE ________

REPORT FOR EXPERIMENT 25

INSTRUCTOR ____________________

Heat of Reaction

A. Hydration of Concentrated Sulfuric Acid

1. Temperature of water. ____________

   Temperature of water after adding acid. ____________

   Temperature change. ____________

2. Is this an endothermic or exothermic reaction? ____________

3. In this reaction, what is the major change, in terms of bonds made or broken, that causes the heat effect?

4. If you had used dilute sulfuric acid rather than concentrated sulfuric acid, would you expect the temperature change to be greater or less?

   Why? ____________

B. Dissolving of Ammonium Chloride

1. Temperature of water. ____________

   Temperature of water after adding salt. ____________

   Temperature change. ____________

2. Is this an endothermic or exothermic reaction? ____________

3. In this reaction, what is the major change, in terms of bonds made or broken, that causes the heat effect?

4. What other heat effect is present in this reaction, acting in the opposite direction?

5. If you had used 40 mL of water and 6 g of ammonium chloride, rather than the 20 mL and 3 g in the experiment, would you expect to get a larger, smaller, or identical temperature change?

Why?   _____

C. Neutralization Reactions

Data Table

		Temp. Before Adding NaOH	Temp. After Adding NaOH	Temp. Change
1. 20.0 mL H$_2$O 10.0 mL HCl 10.0 mL NaOH	Trial 1 Trial 2 Average	_____ _____ _____	_____ _____ _____	_____ _____ _____
2. 20.0 mL H$_2$O 10.0 mL HNO$_3$ 10.0 mL NaOH	Trial 1 Trial 2 Average	_____ _____ _____	_____ _____ _____	_____ _____ _____
3. 20.0 mL H$_2$O 10.0 mL H$_2$SO$_4$ 10.0 mL NaOH	Trial 1 Trial 2 Average	_____ _____ _____	_____ _____ _____	_____ _____ _____
4. 10.0 mL H$_2$O 10.0 mL HCl 10.0 mL NaOH	Trial 1 Trial 2 Average	_____ _____ _____	_____ _____ _____	_____ _____ _____
5. 10.0 mL H$_2$O 10.0 mL HCl 20.0 mL NaOH	Trial 1 Trial 2 Average	_____ _____ _____	_____ _____ _____	_____ _____ _____

(Part C continued)

QUESTIONS AND PROBLEMS

1. (a) What was the average temperature change for neutralization reactions #1, 2, and 3?

    _____

    (b) What do you conclude from this?

2. How many moles of $H_2O(l)$ were formed in each neutralization reaction (#1, 2, and 3)?
    Show calculation set-ups:

    _____

3. Using the average temperature change for reactions #1, 2, 3, how many joules of heat were released during the neutralization reactions?
    Show calculation set-ups:

    _____

4. Using the average joules of heat calculated for these reactions and the moles H_2O produced for each trial, calculate the kJ/mole of H_2O formed.

    _____

5. Calculate the percent error for the experimental heat of neutralization?

    _____

6. (a) How does the average temperature change in neutralization #4 compare to the average change determined for #1, 2, and 3?

    (b) How do you explain this when there was twice as much $HCl(aq)$ involved in #4 vs. #1, 2, and 3?

7. (a) How does the average temperature change for neutralization #5 compare with all the others?

   (b) Explain your answer.

8. What is the kJ/mol H_2O formed for neutralization #5? Support your answer with calculation setups.

EXPERIMENT 26

Distillation of Volatile Liquids

MATERIALS AND EQUIPMENT

Liquids: Distilled water, ethanol (denatured). **Solutions:** ethanol/water(50/50%), red wine.

Special equipment: Distillation flask, 125 or 250 mL; thermometer, 100°C; condenser with bent adapter, heat source without open flame (hot plate or heating mantle with rheostat), boiling stones, and pot holders or mitts.

DISCUSSION

A. Background

The most commonly used method for purification of liquids is distillation, a process by which one liquid can be separated from another liquid or from a nonvolatile solid. There are many applications of distillation. For example, in many arid regions, distillation is used to obtain potable water from seawater. In the petroleum industry, distillation is used to separate the components of crude oil into fractions such as gasoline, kerosene, and diesel fuel.

Evaporation is the escape of molecules from the liquid state to the vapor (gaseous) state.

$$\text{liquid} \xrightarrow{\text{evaporation}} \text{vapor}$$

The rate of evaporation is dependent on the temperature and vapor pressure of the liquid. The vapor pressure of a liquid is the pressure exerted by a vapor in equilibrium with its liquid. When a liquid is heated, its vapor pressure increases until it equals the pressure of the atmosphere above it, at which point the liquid begins to boil. The normal boiling point (bp) of a liquid is the temperature at which the liquid boils when the atmospheric pressure is 1 atm (760 torr), the average atmospheric pressure at sea level.

> **NOTE:** The temperatures in this discussion and procedure refer to the normal boiling points at 1 atm pressure. If the atmospheric pressure at your laboratory is lower than 1 atm, experimental temperatures will be lower than those given in this experiment.

A volatile liquid has relatively weak intermolecular attractions. When such a liquid is placed in an open dish, the liquid evaporates and the vapor diffuses into the atmosphere. During a distillation, the vapor forms in the distilling flask and is condensed back to a liquid and collected in a separate vessel. Thus, in distillation

$$\text{liquid} \xrightarrow[\text{heat}]{\text{evaporation}} \text{vapor} \xrightarrow{\text{condensation}} \text{liquid}$$

Consider the three liquids, ethyl ether (bp 34.5°C), ethanol (bp 78.5°C), and water (bp 100°C). Ethyl ether is more volatile than ethanol, which is more volatile than water. This means that molecules of ethyl ether have less intermolecular attraction than the molecules of the other two compounds. It also means that ethyl ether has a higher vapor pressure than the other two compounds at any temperature up to its boiling point. All three substances boil at atmospheric pressure, but at different temperatures.

When a solution of two volatile liquids is heated in the distilling apparatus shown in Figure 26.1, both compounds are present in the vapor. In an efficient distilling apparatus, the vapor condenses to a liquid and revaporizes many times, continually becoming richer in the lower boiling component. When the vapor reaches the thermometer and passes into the condenser, it is possible to have separated one compound from the other. The apparatus in Figure 26.1 is a simple distillation setup that is not very efficient for separating two liquids unless there is a large difference in their boiling points. However, it serves the purpose for demonstrating the principles of distillation.

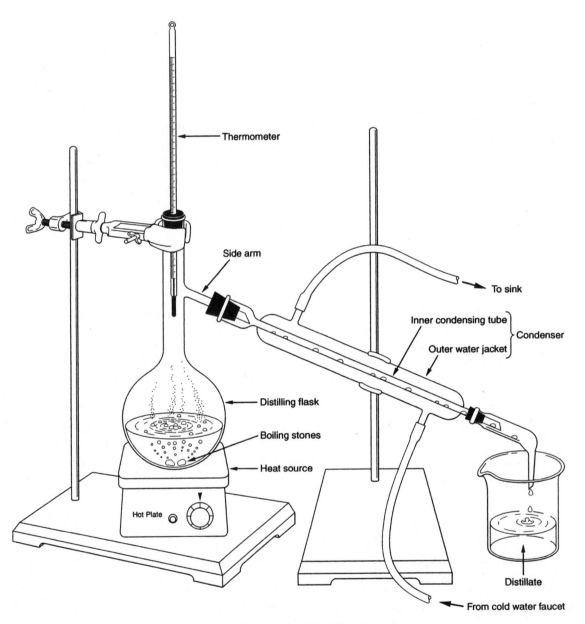

Figure 26.1   A simple distillation setup

As a liquid boils, hot vapor moves up the distilling flask, past the thermometer measuring its temperature, and through the sidearm into the condenser. If the vapor cools off before it reaches the thermometer, it condenses on the sides of the flask and does not reach the thermometer or the condenser. The condenser consists of an inner tube surrounded by a water jacket for cooling the vapor. The vapor loses heat to the flowing water and condenses back to a liquid on the cool surface of the inner tube. The condensed liquid, called the **distillate,** flows down the inner tube and is collected in a beaker or a flask.

Distillation should be conducted slowly and steadily and at such a rate that an equilibrium is maintained between the vapor and the condensate (liquid) where the thermometer is located. A drop of condensate should always appear on the bulb of the thermometer while vapor is flowing into the condenser. **The distilling flask should not be heated to dryness.**

B.   Boiling Points and Heating Curves.

During the first part of this experiment you will observe the boiling points and obtain data to plot heating curves for two liquids, ethanol and water. Each liquid is heated separately in a distilling flask and the temperature is recorded at intervals of 0.5 minute for three minutes and at 1-minute intervals thereafter. A temperature plateau occurs even as heat continues to be added to the system. The time/temperature data, when plotted as a graph, is known as a heating curve. The experimental boiling point is at the plateau where the temperature remains constant over a narrow range of $1° - 2°C$ or less.

C.   Distillation of Solutions Containing Two Volatile Liquids.

In the second part of this experiment, the liquids being distilled are solutions of alcohol and water. When there is more than one volatile liquid in a solution, the distillation is more complex. In theory, the lower-boiling, more volatile liquid will distill first, followed by the higher boiling liquid. Any nonvolatile components will remain in the distilling flask. The theoretical heating curve for two liquids with boiling points of 80°C and 110°C should look like Figure 26.2.

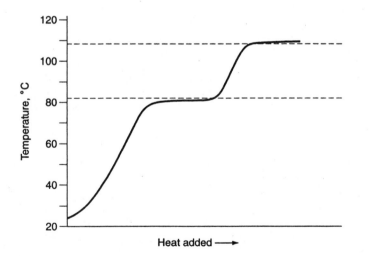

Figure 26.2   Theoretical heating curve for two volatile liquids

In reality, it is difficult to get the two distinct plateaus necessary to separate the two volatile liquids. When you distill a solution of two volatile liquids, the vapor will contain both compounds over much of the temperature range. Due to their relative vapor pressures at the lower temperatures, the vapor will contain a higher concentration of the lower boiling (higher vapor pressure) liquid and the distillate will be mostly the lower boiling component. Later, at the higher end of the boiling range, the vapor will have more of the less volatile compound and the distillate will be mostly the higher boiling component. Thus, in a simple distillation setup such as the one we are using, the temperature plateaus may not be as flat as when distilling pure compounds. In fact, depending on the concentration of the solution and the rate of heating, you may not observe much of a plateau at all.

PROCEDURE

Wear protective glasses.

1. Set the heat source aside and set up the rest of the apparatus as shown in Figure 26.1. For the first distillation you must start with a dry distilling flask. Because the heat source cannot be an open flame, you will use a hot plate or heating mantle. Your instructor will explain the use of the heat source available to you. Set the dial as instructed, keep it away from the flask, and allow it to preheat for five minutes.

 From this point on, remember that the heat source is hot and can burn you when you touch its surface.

2. Every team of students will do three distillations: ethanol, distilled water, and one of the two alcohol/water solutions. Assignments will be made by the instructor.

    a. ethanol   } every team will do

    b. deionized water   } these two distillations

    c. 50/50 ethanol/water solution   } every team will do

    d. alcoholic beverage, red wine   } one of these distillations

First Distillation: Ethanol

3. Fill a dry graduated cylinder with 50 ml of ethanol. Use a funnel to pour the liquid from the graduated cylinder into the distilling flask. Be very careful to avoid pouring the liquid into the side arm. Add two boiling stones. Boiling stones are used to ensure uniform boiling and to avoid superheating.

4. Put the stopper into the distilling flask and position the thermometer just below the side arm. Make sure the apparatus is put together securely and place a beaker or an Erlenmeyer flask under the receiving end of the condenser to collect the distillate. The heat source is not under the distilling flask yet.

5. Attach the bottom hose on the condenser to a cold water spigot and put the top hose into a sink. *Carefully* turn on the spigot to pass water through the condenser.

6. Before beginning the distillation, work out plans with a partner to observe and record data. Once the heating begins, you need to be very observant. One person should watch the thermometer bulb, the temperature, and the clock, and call out the temperatures, while the other should record temperatures that are being called out. Record all temperatures to the nearest 0.1°C.

Remember: Use a pot holder or insulated mitt to prevent injury when moving the heat source.

7. Position the preheated heat source so it is touching the bottom of the distilling flask. Observe the temperature and when the thermometer reads about 34°C, record this temperature for time 0.0 minutes. As you wait for the temperature to reach this point, notice that bubbles begin to emerge from the boiling stones. These bubbles will get more vigorous as the temperature of the liquid increases. Remember, however, the thermometer is not in the liquid. It is measuring the temperature near the side arm, relatively far away from the heat source and the liquid.

8. Continue heating and recording the temperature at 30-second intervals for three minutes and then at 1-minute intervals. Be very alert because sometimes the temperature can shoot up very suddenly once the hot vapor reaches the top of the flask where the thermometer is located. Be very patient because at other times the temperature can change very slowly.

9. Pay attention to the formation of the first drop of liquid on the thermometer. When this drop appears, the liquid-vapor equilibrium has been established. If this drop of liquid evaporates completely, the equilibrium has been lost. Make a note on the data table and adjust the heat source. The heat will need to be turned down slightly if it is too hot near the thermometer bulb for condensation to occur. (You will see the drops of liquid dripping from the side arm as vapor enters the condenser.) The heat will need to be turned up slightly if the vapor is condensing before it reaches the thermometer bulb.

10. Continue recording the temperature for fifteen minutes. When the temperature remains constant for five minutes before the full fifteen minutes have elapsed you may assume that it will remain constant for the remaining time. If the temperature has not stabilized after fifteen minutes, add more time on the data table and continue recording. Turn off the hot plate and separate the distilling flask from the condenser using pot holders to avoid burning yourself. Run cold water over the round bottom of the flask to cool it and the liquid inside.

 11. Empty the ethanol remaining in the distilling flask and the distillate into the ethanol recycling bottle provided. Throw the boiling stones in the waste basket.

Second Distillation: Water

12. Rinse the distillation flask and thermometer with distilled water and add 50 mL of distilled water and two new boiling stones to the flask. Repeat steps 7–10 again. Ask your instructor if it is necessary to increase the temperature of the heat source.

 Pour the distillate and water remaining in the distilling flask into the sink and the boiling stones into the waste basket.

Third Distillation: A Solution of Ethanol and Water

13. Use a cooled distilling flask as before. Add two boiling stones and 50 mL of the ethanol/water solution that has been assigned to your group. Before you begin to heat, note that the time intervals on the data table for this distillation are different. Begin recording at 0.0 minutes when the temperature reaches about 34°C. You will record at 1-minute intervals during the first six minutes and at 2-minute intervals thereafter. It may take about 45 minutes to do this distillation. Your instructor may choose to terminate these distillations before the final plateau is reached in order to save time. If the drop of liquid on the thermometer

disappears before the boiling point of water is reached, turn up the heat slightly. Stop recording after 45 minutes. When the temperature reaches the boiling point of water and remains there for 2–4 minutes, you may stop the distillation sooner than 45 minutes.

Turn off and unplug the heat source. Use the pot holders to separate the distilling flask from the condenser and the heat source.

WASTE DISPOSE OF PROPERLY

Empty the water/alcohol mixtures into the sink and flush with water. Throw the boiling stones into the waste basket.

14. Exchange data for the third distillation with another team that distilled the alternate solution.

15. Complete two graphs using time as the independent variable and temperature as the dependent variable. Use the graph paper provided or make a computer graph. Review the instructions in Study Aid 3 if necessary.

Graph 1. Plot temperature vs. time data for distillations 1 and 2.

Graph 2. Plot temperature vs. time data for the other two distillations (solutions).

REPORT FOR EXPERIMENT 26

Distillation of Volatile Liquids

Data Table

time, minutes	temp, °C ethanol	temp, °C H_2O		time, minutes	temp, °C 50% ethanol	temp, °C red wine	
0.0				0.0			
0.5				1.0			
1.0				2.0			
1.5				3.0			
2.0				4.0			
2.5				5.0			
3.0				6.0			
4.0				8.0			
5.0				10.0			
6.0				12.0			
7.0				14.0			
8.0				16.0			
9.0				18.0			
10.0				20.0			
11.0				22.0			
12.0				24.0			
13.0				26.0			
14.0				28.0			
15.0				30.0			
				32.0			
				34.0			
				36.0			
				38.0			
				40.0			
				45.0			

QUESTIONS AND PROBLEMS

1. The boiling points of three liquids are provided below.

   Acetone, bp = 56.2°C   Methanol, bp = 65°C   Ethylene glycol, bp = 198°C

   Which liquid is the least volatile? (circle your choice)

   Which liquid has the weakest attractive forces between its molecules? (underline your choice)

2. If the drop of liquid on the thermometer disappears while the vapor is visibly condensing in the side arm, which process is proceeding faster in the flask, evaporation or condensation? (Circle your choice) How would you get a drop of liquid back?

3. Define boiling point.

4. Why is the thermometer bulb placed above the liquid beside the side arm of the distilling flask and not in the liquid?

5. What is the purpose of

   (a) the inner tube of the condenser?

   (b) the outer jacket of the condenser?

6. When the red wine is distilled, why does the distillate remain colorless throughout the full temperature range?

7. Explain *briefly* how distillation can be used to purify seawater to make salt-free water.

8. One team collected distillate from the wine in four 10 mL fractions in a graduated cylinder. What is the difference between the composition of the first 10 mL collected and the last 10 mL?

   Explain this difference.

Use your graphs to answer the following questions:

9. (a) What is the experimental boiling point of ethanol? _____ water? _____

   (b) What is the boiling point range for the 50% ethanol solution? _____

10. Why did it take so much longer to reach the boiling point of water in the solutions of ethanol and water than in the pure water?

Graph 1

Graph 2

EXPERIMENT 27

Organic Chemistry—Hydrocarbons

MATERIALS AND EQUIPMENT

Solids: calcium carbide (about 3/8 in. lumps) (CaC_2). **Liquids:** pentene (amylene) (C_5H_{10}), heptane (C_7H_{16}), kerosene, and toluene ($C_6H_5CH_3$). **Solutions:** 5% bromine (Br_2) in 1,1,1-trichloroethane (CCl_3CH_3) and 0.1 M potassium permanganate ($KMnO_4$). Wooden splints.

See Study Aid 6 for a brief introduction to Organic Chemistry.

Hydrocarbons

Hydrocarbons are organic compounds made up entirely of carbon and hydrogen atoms. Their principal natural sources are coal, petroleum, and natural gas. Hydrocarbons are grouped into several series by similarity of molecular structure. Some of these are the alkanes, alkenes, alkynes, and aromatic hydrocarbons.

Alkanes

Also known as the paraffins or **saturated hydrocarbons,** the **alkanes** are straight- or branched-chain hydrocarbons having only single bonds between carbon atoms. They are called saturated hydrocarbons because all their carbon-carbon bonds are single bonds.

The first 10 members of the alkane series and their molecular formulas are listed below:

Methane CH_4	Hexane C_6H_{14}
Ethane C_2H_6	Heptane C_7H_{16}
Propane C_3H_8	Octane C_8H_{18}
Butane C_4H_{10}	Nonane C_9H_{20}
Pentane C_5H_{12}	Decane $C_{10}H_{22}$

Like most organic substances, the alkanes are combustible. The products of their complete combustion are carbon dioxide and water. The reactions of the alkanes are of the substitution type; that is, some atom or group of atoms is substituted for one or more of the hydrogen atoms in the hydrocarbon molecule. For example, in the bromination of methane, a bromine atom is substituted for a hydrogen atom. This reaction does not occur appreciably in the dark at room temperature but is catalyzed by ultraviolet light. The equation is

$$CH_4 + Br_2 \xrightarrow[\text{light}]{\text{Ultraviolet}} CH_3Br + HBr$$

Methane   Methyl bromide
(Bromethane)

The equation for the combustion of propane is

$$CH_3CH_2CH_3 + 5O_2 \xrightarrow{\Delta} 3CO_2 + 4H_2O$$

Alkenes

Also known as the olefins, the alkenes are a series of straight- or branched-chain hydrocarbons containing a carbon-carbon double bond in their structures. They are considered to be unsaturated hydrocarbons The first two members of the series are ethene (C_2H_4) and propene (C_3H_6). The general formula for alkenes is C_nH_{2n}. Their structural and condensed structural formulas are:

Ethene (ethylene)   Propene (propylene)

The functional group of this series is the carbon-carbon double bond ($C=C$); it is a point of high reactivity. Alkenes undergo addition-type reactions; that is, other groups are added to the double bond, causing the molecule to become saturated. For example, when hydrogen is added, one H atom from H_2 is added to each carbon atom of the double bond to saturate the molecule, forming an alkane:

$$CH_3CH=CH_2 + H_2 \xrightarrow[\text{Heat and Pressure}]{\text{Ni Catalyst}} CH_3CH_2CH_3$$

Propene   Propane

When a halogen such as bromine is added, one Br atom from Br_2 is added to each carbon atom of the double bond to saturate the molecule:

1,2-Dibromoethane
(Ethylene dibromide)

Evidence that bromine has reacted is the disappearance of the red-brown color of free bromine. Other reactions of olefins also show the increased reactivity of the alkenes over the alkanes.

Unsaturated hydrocarbons can be oxidized by potassium permanganate. The reaction is known as the **Baeyer test for unsaturation**. Evidence that reaction has occurred is the rapid disappearance (within a few seconds) of the purple color of the permanganate ion. The resulting reaction products will not be colorless. Potassium permanganate is a very strong oxidizing agent and gives similar results when reacted with other oxidizable substances, such as alcohols.

Alkynes

Also called the **acetylenes,** the alkynes are another class of **unsaturated hydrocarbons,** but they contain a carbon-carbon triple bond in their structures. The first two members of this series are acetylene (ethyne) and propyne:

$$H-C\equiv C-H \qquad\qquad CH_3C\equiv CH$$

Ethyne (Acetylene)   Propyne

Acetylene is the most important member of this series and can be prepared from calcium carbide and water. The equation for this reaction is

$$CaC_2(s) + 2H_2O(l) \longrightarrow CH{\equiv}CH(g) + Ca(OH)_2(aq)$$

Mixtures of acetylene and air are explosive. The alkynes undergo addition-type reactions similar to those of the alkenes.

Aromatic Hydrocarbons

The parent substance of this class of hydrocarbons is benzene (C_6H_6). From its formula benzene appears to be a highly unsaturated molecule; the corresponding six-carbon alkane contains 14 hydrogen atoms per molecule (C_6H_{14}). However, the chemical reactions of benzene show that its behavior is like that of the saturated hydrocarbons in many respects. Its reactions are primarily of the substitution type. In the past, benzene was used extensively in student laboratories to illustrate the properties of aromatic hydrocarbons. Within the last decade, studies have shown benzene to be a cancer-causing substance and it is being eliminated from many experiments. We will use toluene in this experiment instead of benzene.

The carbon atoms in a benzene molecule are arranged in a six-membered ring structure, with one hydrogen atom bonded to each carbon atom. The following diagrams represent the benzene molecule; in the second and third structures it is understood that a carbon and a hydrogen atom are present at each corner of the hexagon or benzene ring.

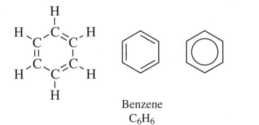

Benzene
C_6H_6

Toluene
$C_6H_5CH_3$

PROCEDURE

Wear protective glasses.

Record your observation as you do each experiment.

In the reactions below, heptane will be used to represent the saturated hydrocarbons; pentene (amylene), the unsaturated hydrocarbons; and toluene, the aromatic hydrocarbons.

> ⚠️ Hydrocarbons are extremely flammable and should not be handled near open flames. Avoid inhaling the vapors, contact with skin and clothing, and do not ingest.

> **WASTE** DISPOSE OF PROPERLY Dispose of waste and reaction products in the "organic solvent" waste container provided. The **WASTE** DISPOSE OF PROPERLY logo will be used to remind you to dispose those wastes properly.

A. Combustion

Obtain about 1 mL (no more) of heptane in an evaporating dish and start it burning by carefully bringing a lighted match or splint to it. Repeat with an equally small volume of pentene. Note the characteristics of the flames.

B. Reaction with Bromine

⚠️ **CAUTION:** Dispense bromine solution under the hood, and be especially careful not to spill bromine on your hands.

Take three clean dry test tubes. Place about 1 mL of heptane in the first tube and 1 mL of pentene in the second, amd 1 mL of toluene in the third. Add 3 drops of 5 percent bromine in trichloroethane solution to each sample; stopper the tubes and note the results. Any tube that still shows bromine color after 1 minute should be exposed to sunlight or to a strong electric light for an additional 2 minutes. **[WASTE DISPOSE OF PROPERLY]**

C. Reaction with Potassium Permanganate

⚠️ The Baeyer test for unsaturation in hydrocarbons involves the reaction of hydrocarbons with potassium permanganate solution. Evidence that reaction has occurred is the rapid disappearance (within a few seconds) of the purple color of the permanganate ion. Potassium permanganate is a very strong oxidizing agent and gives similar results when reacted with other oxidizable substances, such as alcohols.

Add 2 drops of potassium permanganate solution to about 1 mL each of heptane, pentene, and toluene in test tubes. Mix and note the results.

D. Kerosene

Determine which class of hydrocarbons (alkanes, alkenes, or aromatic) kerosene belongs to by reacting it with bromine and with potassium permanganate, as in Tests B and C. **[WASTE DISPOSE OF PROPERLY]**

E. Acetylene

In this part of the experiment you will prepare acetylene and test its combustibility.

Fill a 400 mL beaker nearly full of tap water. Fill three test tubes (18 × 150 mm) with water as follows: Tube (1) completely full; Tube (2) 15 mL; and Tube (3) 6 mL.

Obtain a small lump of calcium carbide from the reagent bottle and drop it into the beaker of water. (See Figure 27.1.) Place your thumb over the full test tube of water (Tube 1) and invert it in the beaker. Hold the tube over the bubbling acetylene and, when it is full of gas, stopper it while the tube is still under the water. Displace the water in the other two tubes in the same manner; **stopper them immediately after the water is displaced.**

Test the contents of each tube as follows:

Tube 1. Bring the mouth of the tube to the burner flame as you remove the stopper. After the acetylene ignites, tilt the mouth of the tube up and down.

Tube 2. Bring the mouth of the tube to the burner flame as you remove the stopper.

  Tube 3. Wrap the tube in a towel and bring the mouth of the tube to the burner flame as you remove the stopper. This sample is the most highly explosive of the three samples tested.

F. Solubility Tests

Test the solubility of heptane, pentene, and toluene in water by adding 1 mL (or less) of each hydrocarbon to about 5 mL portions of water in a test tube. Shake each mixture for a few seconds and note whether they are soluble. For any that are not soluble note the relative density of the hydrocarbon with respect to water.

Test the miscibility of these three hydrocarbons with each other by mixing about 1 mL of each in a **dry** test tube.

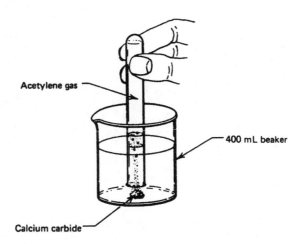

Figure 27.1   Collecting acetylene from calcium carbide-water reaction

NAME ______________________________

SECTION __________ DATE __________

REPORT FOR EXPERIMENT 27

INSTRUCTOR ____________________

Hydrocarbons

A. Combustion

1. Describe the combustion characteristics of heptane and pentene.

2. (a) Write a balanced equation to represent the complete combustion of heptane.

   (b) How many moles of oxygen are needed for the combustion of 1 mole of heptane in this reaction?

   ____

B. and C. Reaction with Bromine and Potassium Permanganate

Data Table: Place an X in the column where a reaction was observed.

	Heptane (Saturated Hydrocarbon)	**Pentene (Unsaturated Hydrocarbon)**	**Toluene (Aromatic Hydrocarbon)**
Immediate reaction with Br_2 (*without* exposure to *light*)			
Slow reaction with Br_2 (*or only after exposure to light*)			
Reaction with $KMnO_4$			

1. Which of the thre hydrocarbons reacted with bromine (without exposure to light)?

   ____

2. Write an equation to illustrate how heptane reacts with bromine when the reaction mixture is exposed to sunlight.

3. Write an equation to illustrate how pentene reacts with bromine. Assume the pentene is $CH_3CH_2CH=CHCH_3$ and use structural formulas.

4. Which of the three hydrocarbons tested gave a positive Baeyer test?

_____

D. Kerosene

1. (a) Did you observe any evidence of reaction with bromine before exposure to light? If so, describe.

   (b) Did you observe any evidence of reaction with bromine after exposure to light? If so, describe.

2. Did you observe any evidence of reaction with potassium permanganate? If so, describe.

3. Based on these tests (bromine and potassium permanganate), to which class of hydrocarbon does kerosene belong?

_____

E. Acetylene

1. Describe the combustion characteristics of acetylene:

   (a) Tube 1.

   (b) Tube 2.

   (c) Tube 3.

2. Write an equation for the complete combustion of acetylene.

F.   Solubility Tests

1. Which of the three hydrocarbons tested are soluble in water?

2. From your observations what do you conclude about the density of hydrocarbons with respect to water?

QUESTIONS AND PROBLEMS

1. Do you expect that acetylene would react with bromine without exposure to light? Explain your answer.

2. Write condensed structural formulas for the three different isomers of pentane, all having the molecular formula C_5H_{12}. See Study Aid 6 for help if necessary.

3. Write condensed structural formulas for (a) ethene, (b) propene, and (c) the three different isomeric butenes (C_4H_8). See Study Aid 6 for help if necessary.

EXPERIMENT 28

Alcohols, Esters, Aldehydes, and Ketones

MATERIALS AND EQUIPMENT

Solids: Copper wire (No. 18, with spiral); salicylic acid [$C_6H_4(COOH)(OH)$]. **Liquids:** acetic acid, glacial (CH_3COOH); acetone [$(CH_3)_2C{=}O$]; ethyl alcohol (95% C_2H_5OH); isoamyl alcohol ($C_5H_{11}OH$); isopropyl alcohol (iso-C_3H_7OH); methyl alcohol (CH_3OH). **Solutions:** dilute (6 M) ammonium hydroxide (NH_4OH), 10 percent glucose ($C_6H_{12}O_6$), 10 percent formaldehyde ($H_2C{=}O$), 0.1 M potassium permanganate ($KMnO_4$), 0.1 M silver nitrate ($AgNO_3$), 10 percent sodium hydroxide (NaOH), and dilute (3 M) and concentrated sulfuric acid (H_2SO_4).

DISCUSSION

In this experiment we will examine some of the properties and characteristic reactions of four classes of organic compounds: alcohols, esters, aldehydes, and ketones.

Alcohols

The formulas of alcohols may be derived from alkane hydrocarbon formulas by replacing a hydrogen atom with a hydroxyl group (OH). In the resulting alcohols the OH group is bonded to the carbon atom by a covalent bond and is not an ionizable hydroxide group. Examples follow:

Alkane	Alcohol	Name of Alcohol*
CH_4	CH_3OH	Methyl alcohol (Methanol)
CH_3CH_3	CH_3CH_2OH	Ethyl alcohol (Ethanol)
$CH_3CH_2CH_3$	$CH_3CH_2CH_2OH$	n-Propyl alcohol (1-Propanol)
$CH_3CH_2CH_3$	CH_3CHCH_3 $\quad\;\;\,|$ $\quad\;\;OH$	Isopropyl alcohol (2-Propanol)

Thus there is an entire homologous series of alcohols. The functional group of the alcohols is the hydroxyl group, OH.

Esters

This class of organic compounds may be formed by reacting alcohols with organic acids. Esters generally have a pleasant odor; many of them occur naturally, being found mainly in fruits and fatty material.

Methyl acetate will be used as an example illustrating the formation of an ester. When acetic acid and methyl alcohol are reacted together, using sulfuric acid as a catalyst, a molecule of water is split out between a molecule of the acetic acid and a molecule of the alcohol, forming the ester. The equation is

$$\underset{\text{Acetic acid}}{CH_3\overset{\overset{\textstyle O}{\|}}{C}-OH} \;+\; \underset{\text{Methyl alcohol}}{CH_3-OH} \;\xrightarrow[\Delta]{H_2SO_4}\; \underset{\text{Methyl acetate}}{CH_3\overset{\overset{\textstyle O}{\|}}{C}-O-CH_3} \;+\; H_2O$$

The functional group characterizing organic acids is

$$-\overset{\overset{\textstyle O}{\|}}{C}-OH \qquad \text{or} \qquad -COOH$$

It is called a **carboxyl group.**

Esters are named in the following manner. The first part of the name is taken from the name of the alcohol, the second part is derived by adding the suffix ate to the identifying stem of the acid. Thus acetic becomes acetate, and the name of the ester derived from methyl alcohol and acetic acid is methyl acetate.

Aldehydes and Ketones

The functional groups of the aldehydes and ketones are

Alcohol	Aldehyde	Ketone
CH_3OH	$\underset{H-C=O}{\overset{H\;\;\;}{\underset{\|\;\;\;}{}}}$ Methanal (formaldehyde)	—
CH_3CH_2OH	$\underset{CH_3C=O}{\overset{H\;\;\;\;\;}{\underset{\|\;\;\;\;\;}{}}}$ Ethanal (acetaldehyde)	—
$CH_3CH_2CH_2OH$	$\underset{H}{\underset{\|}{CH_3CH_2C=O}}$ Propanal (propionaldehyde)	—
$\underset{OH}{\underset{\|}{CH_3CHCH_3}}$	—	—	$CH_3\overset{\overset{\textstyle}{}}{\underset{\underset{\textstyle O}{\|}}{C}}CH_3$ Acetone (propanone)

Aldehydes and ketones may be obtained by oxidizing alcohols. One major difference between aldehydes and ketones is that aldehydes are very easily oxidized to acids, but ketones are not easily further oxidized. Thus aldehydes are good reducing agents. Chemical reactions for distinguishing aldehydes and ketones are based on this difference.

PROCEDURE

Wear protective glasses.

> Record your observations immediately on the report form as you work through the procedure.
>
> Reagents used in this experiment are highly flammable and several are poisonous. Work cautiously away from heat and open flames, and avoid inhalation of vapors, contact with skin and clothing, and do not ingest.
>
> Dispose of all reagents in the "organic solvent" waste container provided. The symbol ![WASTE] is used throughout this procedure to remind you of this disposal requirement.

A. Combustion of Alcohols

Obtain about 1 mL (no more) of methyl alcohol in an evaporating dish and ignite the alcohol with a match or burning splint. Repeat with equally small volumes of ethyl alcohol and isopropyl alcohol.

B. Oxidation of Alcohols

1. Oxidation with Potassium Permanganate. Mix 3 mL of methyl alcohol with 12 mL of water and divide the solution into three equal portions, placing them in three test tubes. To a fourth tube add 5 mL water. Add 1 drop of 10 percent sodium hydroxide to the first tube and 1 drop of dilute sulfuric acid to the second. Now add 1 drop of potassium permanganate solution to each of the four tubes. Mix and note how long it takes for the reaction to occur in each of the first three tubes, using the fourth tube as a reference tube. Disappearance of the purple permanganate color is evidence of reaction. (Patience, some reactions take a long time.) ![WASTE]

Repeat this oxidation procedure, using isopropyl alcohol instead of methyl alcohol.

2. Oxidation with Copper(II) Oxide. Put about 2 mL of methyl alcohol in a test tube. Obtain from the reagent shelf about a 20 cm piece of copper wire with a four- or five-turn spiral at one end. Warm the alcohol slightly to promote alcohol vapors in the tube. Heat the copper spiral in the hottest part of the burner flame to get a good copper(II) oxide coating. Do not overheat the copper or it will melt. While the copper spiral is very hot, lower it part way into the tube (not to the liquid) and note the results. Heat the wire again and lower it into the tube, finally dropping it into the liquid alcohol. Remove the wire and gently waft the vapors from the tube to your nose to detect the odor of formaldehyde resulting from the oxidation of the methyl alcohol. ![WASTE] **Return the copper wire to the reagent shelf.**

C. Formation of Esters

Take three test tubes and mix the following reagents in them.

Tube 1:   3 mL ethyl alcohol, 0.5 mL glacial acetic acid, and 10 drops of concentrated sulfuric acid.

Tube 2: 3 mL isoamyl alcohol, 0.5 mL glacial acetic acid, and 10 drops concentrated sulfuric acid.

Tube 3: Salicylic acid crystals (about 1 cm deep in the tube), 2 mL methyl alcohol, and 10 drops concentrated sulfuric acid.

Heat the tubes by placing them in boiling water for 3 minutes.

Products formed:

Tube 1: Ethyl acetate.

Tube 2: Isoamyl acetate.

Tube 3: Methyl salicylate.

After heating, pour a small amount of each product onto a piece of filter paper and **carefully** smell it and describe the odor. 〔WASTE DISPOSE OF PROPERLY〕 Dispose of the filter paper in the solid trash.

D. Tollens Test for Aldehydes

This test is based on the ability of the aldehyde group to reduce silver ion in solution, forming either a black deposit of free silver or a silver mirror. The aldehyde group is oxidized to an acid in the reaction. Tollens reagent is made by reacting silver nitrate solution with dilute ammonium hydroxide. Rinse all glass equipment with distilled water before use.

1. Preparation of Tollens Reagent: **Thoroughly** clean three test tubes with soap and water and rinse with distilled water. To 8 mL of 0.1 M silver nitrate solution in one of these tubes add one drop of 10% NaOH to generate a brown precipitate of silver oxide. Now add dilute ammonium hydroxide 1 drop at a time until the brown precipitate of silver oxide that was formed just dissolves (mix after each drop is added). Now add 7 mL of distilled water, mix, and divide the solution (Tollens reagent) equally among the three test tubes.

2. To each of the tubes containing the freshly prepared Tollens reagent, add the following and mix.

Tube 1: 2 drops 10% formaldehyde

Tube 2: 2 drops acetone

Tube 3: 5 drops 10% glucose

Allow the tubes to stand undisturbed and note the results. The solution containing the glucose may take 10 to 15 minutes to react.

〔WASTE DISPOSE OF PROPERLY〕 Dispose of these solutions in the heavy metals waste container.

NAME ____________________

SECTION __________ DATE __________

REPORT FOR EXPERIMENT 28

INSTRUCTOR ____________________

Alcohols, Esters, Aldehydes, and Ketones

A. Combustion of Alcohols

1. Compare the combustion characteristics of methyl, ethyl, and isopropyl alcohols, in terms of color and luminosity of their flames.

2. What type of flame would you predict for the combustion of amyl alcohol ($C_5H_{11}OH$)?

3. Write and balance the equation for the complete combustion of ethyl alcohol.

B. Oxidation of Alcohols

1. **Oxidation with Potassium Permanganate**

(a) Time required for oxidation of methyl alcohol by potassium permanganate:

Tube 1: Alkaline solution. __________

Tube 2: Acid solution. __________

Tube 3: Neutral solution. __________

(b) Time required for oxidation of isopropyl alcohol by potassium permanganate:

Tube 1: Alkaline solution. __________

Tube 2: Acid solution. __________

Tube 3: Neutral solution. __________

(c) Balance the equation for the oxidation of methyl alcohol:

$$CH_3OH + \quad KMnO_4 \longrightarrow H_2C{=}O + \quad KOH + \quad H_2O + \quad MnO_2$$

2. Oxidation with Copper(II) Oxide

(a) Write the equation for the oxidation reaction that occurred on the copper spiral when it was heated.

(b) What evidence of oxidation or reduction did you observe when the heated Cu spiral was lowered into methyl alcohol vapors?

(c) Write and balance the oxidation-reduction equation between methyl alcohol and copper(II) oxide.

C.   Formation of Esters

1. Describe the odor of:

   (a) Ethyl acetate.

   (b) Isoamyl acetate.

   (c) Methyl salicylate.

2. Write an equation to illustrate the formation of ethyl acetate from ethyl alcohol and acetic acid.

3. (a) The formula for isoamyl alcohol is $CH_3CH(CH_3)CH_2CH_2OH$. Write the formula for isoamyl acetate.

(b) The formula for salicylic acid is

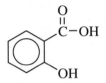

Write the formula for methyl salicylate.

D.   Tollens Test for Aldehydes

1. How is a positive Tollens test recognized?

2. Which of the substances tested gave a positive Tollens test?

3. Circle the formula(s) of the compounds listed that will give a positive Tollens test:

$$CH_3OH \qquad C_2H_5OH \qquad CH_3\overset{H}{\underset{}{C}}{=}O \qquad CH_3\underset{\underset{O}{\|}}{C}CH_3 \qquad Na_2CO_3$$

4. Write the formula for the oxidation product formed from formaldehyde in the Tollens test.

QUESTIONS AND PROBLEMS

1. There are four butyl alcohols of formula C_4H_9OH. Write their condensed structural formulas.

2. Write the name of the ester that can be derived from the following pairs of acids and alcohols:

Alcohol	Acid	Ester
Methyl alcohol	Acetic acid	
Ethyl alcohol	Formic acid	
Isopropyl alcohol	Butyric acid	

3. Write condensed structural formulas for all the aldehydes and ketones containing five carbon atoms. The molecular formula is $C_5H_{10}O$.

STUDY AID I

Significant Figures

Every measurement that we make has some inherent error due to the limitations of the measuring instrument and the experimenter. The numerical value recorded for a measurement should give some indication of the reliability (precision) of that measurement. In measuring a temperature using a thermometer calibrated at one-degree intervals we can easily read the thermometer to the nearest one degree, but we normally estimate and record the temperature to the nearest tenth of a degree (0.1°C). For example, a temperature falling between 23°C and 24°C might be estimated at 23.4°C. There is some uncertainly about the last digit, 4, but an estimate of it is better information than simply reporting 23°C or 24°C. If we read the thermometer as "exactly" twenty-three degrees, the temperature should be reported as 23.0°C, not 23°C, because 23.0°C indicates our estimate to the nearest 0.1°C. Thus in recording any measurement, we retain one uncertain digit. The digits retained in a physical measurement are said to be significant, and are called **significant figures.**

Some numbers are exact (have no uncertain digits) and therefore have an infinite number of significant figures. Exact numbers occur in simple counting operations, such as 5 bricks, and in defined relationships, such as 100 cm = 1 meter, 24 hours = 1 day, etc. Because of their infinite number of significant figures, exact numbers do not limit or determine the number of significant figures in a calculation.

Counting Significant Figures. Digits other than zero are always significant. Depending on their position in the number, zeros may or may not be significant. There are several possible situations:

1. All zeros between other digits in a number are significant; for example: 3.076, 4002, 790.2. Each of these numbers has four significant figures.

2. Zeros to the left of the first nonzero digit are used to locate the decimal point and are not significant. Thus 0.013 has only two significant figures (1 and 3); The zero is not significant.

3. Zeros to the right of the last nonzero digit, and to the right of the decimal point are significant, for they would not have been included except to express precision. For example, 3.070 has four significant figures; 0.070 has two significant figures.

4. Zeros to the right of the last nonzero digit, but to the left of the decimal, as in the numbers 100, 580, 37000, etc., may not be significant. For example, in 37000 the measurement might be good to the nearest 1000, 100, 10, or 1. There are two conventions which may be used to show the intended precision. If all the zeros are significant, then an expressed decimal may be added, as 580., or 37000. But a better system, and one which is applicable to the case when some but not all of the zeros are significant, is to express the number in exponential notation, including only the significant zeros. Thus for 300, if the zero following 3 is significant, we would write 3.0×10^2. For 17000, if two zeros are significant, we would write 1.700×10^4. The number we correctly expressed as 580. can also be expressed as 5.80×10^2. With exponential notation there is no doubt as to the number of significant figures.

Addition or Subtraction. The result of an addition or subtraction should contain no more digits to the right of the decimal point than are in that quantity which has the least number of digits to the right of the decimal point. Perform the operation indicated and then round off the number to the proper significant figure.

Example: 24.372
             72.21
              6.1488
            ‾‾‾‾‾‾‾‾
            102.7308 (102.73)

Since the digit 1 in 72.21 is uncertain, the sum can have no digits beyond this point, so the sum should be rounded off to 102.73.

Multiplication or Division. In multiplication or division, the answer can have no more significant figures than the factor with the least number of significant figures. In multiplication or division, the position of the decimal point has nothing to do with the number of significant figures in the answer.

Example: $3.1416 \times 7.5 \times 252 = 5937.624(5.9 \times 10^3)$

The operations of arithmetic supply all the digits shown, but this does not make the answer precise to seven significant figures. Most of these digits are not realistic because of the limited precision of the number 7.5. So the answer must be rounded to two significant figures, 5900 or 5.9×10^3. It should be emphasized that in rounding-off the number you are not sacrificing precision, since the digits discarded are not really meaningful.

Example: $\dfrac{(27.52)(62.5)}{1.22} = 1409.836(1.41 \times 10^3)$

The answer should contain three (3) significant figures.

STUDY AID 2

Formulas and Chemical Equations

A.   Formulas for Ionic Compounds

Formulas are combinations of chemical **symbols** which represent molecular and ionic compounds. If the compound is ionic it is composed of ions and the charges on the ions must add up to zero in the formula for the compound.

Chemical **equations** are combinations of formulas which represent chemical reactions. Thus, in order to write correct chemical equations, it is necessary to know how to write correct formulas.

Writing formulas for ionic compounds requires combining ions so that the compound has a net charge of zero. The positive ion is usually written first. **Ions** are atoms or groups of atoms that have either a positive or a negative electrical charge. The relative electrical charge on an ion is indicated by a plus ($+$) or minus ($-$) sign together with a number. The number and the $+$ or $-$ sign are written as a superscript at the upper right corner of the symbol for the ion. For example, Na^+ represents a sodium ion with a $+1$ charge (the 1 is omitted but understood to be present); S^{2-} represents a sulfide ion with a -2 charge. **Polyatomic ions** are groups of atoms with a collective charge. SO_4^{2-} represents a sulfate ion with a -2 charge and is composed of one sulfur atom and four oxygen atoms.

Because all compounds have a net charge of zero, the compound sodium chloride is composed of the sodium ion ($1+$) and the chloride ion ($1-$) combined to form $NaCl$ because $1+$ and $1-$ add up to zero. When more than one atom of an element is used in the fomula of a compound, it is indicated by a numerical subscript written to the right of the element: eg., $CaCl_2$. This formula indicates one Ca^{2+} ion and two Cl^- ions.

The subscripts are determined by combining the smallest number of ions to make the sum of the charges equal to zero. To write the formula for the compound made up of Ba^{2+} and P^{3-}, the lowest common multiple of the charges is 6. Therefore, we need to combine three $2+$ ions and two $3-$ ions $[3(2 +) + 2(3-) = 0]$ and the formula for the compound is Ba_3P_2 (barium phosphide).

If an ion is polyatomic and two or more of these ions are needed to make the sum of the charges equal zero, the ion is enclosed in parentheses and the subscript is written after the parenthesis; e.g. $Cu(NO_3)_2$. This formula indicates one Cu^{2+} ion and two NO_3^- ions.

In the following examples, the ions shown are used to illustrate the formation of possible compounds. The first series is hypothetical but will be used to illustrate the principal of combining positive and negative ions so the net charge is zero.

Hypothetical Ions	Ion and Polyatomic Ion Examples*
X^+	Na^+, K^+, NH_4^+
Y^{2+}	$Ca^{2+}, Pb^{2+}, Mg^{2+}$
Z^{3+}	Al^{3+}, Fe^{3+}
E^-	Br^-, Cl^-, NO_3^-, OH^-
G^{2-}	$O^{2-}, S^{2-}, CrO_4{}^{2-}, SO_4{}^{2-}$
J^{3-}	$P^{3-}, PO_4{}^{3-}, BO_3{}^{3-}$

*A list of more common ions is provided at the end of this manual

Examples of compounds formed by combining positive and negative ions:

Hypothetical Ions	Net Charge Calculation	Hypothetical Compounds	Specific Compounds with Corresponding Ion Ratios
X^+, E^-	$(1+) + (1-) = 0$	XE	$NaCl, NH_4Cl, KBr, NaNO_3$
Y^{2+}, E^-	$(2+) + 2(1-) = 0$	YE_2	$CaCl_2, Mg(NO_3)_2, Ba(OH)_2$
X^+, G^{2-}	$2(1+) + (2-) = 0$	X_2G	$Na_2O, (NH_4)_2SO_4, K_2CrO_4$
Z^{3+}, E^-	$(3+) + 3(1-) = 0$	ZE_3	$AlCl_3, Fe(NO_3)_3$
Z^{3+}, G^{2-}	$2(3+) + 3(2-) = 0$	Z_2G_3	$Al_2O_3, Fe_2(CO_3)_3$
Y^{2+}, J^{3-}	$3(2+) + 2(3-) = 0$	Y_3J_2	$Ca_3(PO_4)_2, Cu_3(BO_3)_2$
Y^{2+}, G^{2-}	$(2+) + (2-) = 0$	YG	$MgO, CaSO_4, PbCrO_4$

B. Writing Equations

A chemical change or reaction results in the formation of products whose compositions are different from the starting substances (reactants). A **chemical equation** is a shorthand expression for a chemical reaction. Substances in the equation are represented by their formulas. The equation indicates both the reactants and their products. The reactants are written on the left side and the products on the right side of the equation. An arrow ($\longrightarrow$) pointing to the products separates the reactants from the products. A plus sign ($+$) is used to separate one reactant (or product) from another.

Reactants $\longrightarrow$ Products

Often, a phase label is added after each formula to indicate the phase of the reactant or product represented by the formula. The phase labels are: (s) = solid; (l) = liquid; (g) = gas; (aq) = aqueous.

Example 1. The combustion of magnesium in oxygen or air.

Word equation: Magnesium + Oxygen ⟶ Magnesium oxide

Formula equation: $Mg + O_2 \longrightarrow MgO$ (unbalanced)

Phase labels included: $Mg(s) + O_2(g) \longrightarrow MgO(s)$ (unbalanced)

Example 2. Barium chloride and sulfuric acid solutions are mixed

Word equation: Barium chloride + sulfuric acid ⟶
barium sulfate + hydrochloric acid

Formula equation: $BaCl_2 + H_2SO_4 \longrightarrow BaSO_4 + HCl$ (unbalanced)

Phase labels included: $BaCl_2(aq) + H_2SO_4(aq) \longrightarrow BaSO_4(s) + HCl(aq)$
(unbalanced)

Example 3. Hydrochloric acid is added to sodium carbonate

Word equation: Sodium carbonate + hydrochloric acid ⟶
sodium chloride + carbon dioxide + water

Formula equation: $Na_2CO_3 + HCl \longrightarrow NaCl + CO_2 + H_2O$ (unbalanced)

Phase labels included: $Na_2CO_3(s) + HCl(aq) \longrightarrow NaCl(aq) + CO_2(g) + H_2O(l)$
(unbalanced)

C. Balancing Equations

There is no detectable change in mass resulting from a chemical reaction. Therefore, the mass of the products must equal the mass of the reactants before the chemical change occurred. In representing the chemical change by an equation, this conservation of mass is attained by **"balancing the equation"**. Because each atom has a particular mass, we balance an equation by adjusting the number of atoms of each kind of element to be the same on each side of the equation. This adjustment is never made by changing subscripts in correct formulas. The adjustment is always made by adding small whole numbers (coefficients) in front of the formulas to adjust the number of atoms in the reactants and products as needed.

If the first equation above is written and left as is ($Mg + O_2 \longrightarrow MgO$), it is not balanced since there are two oxygen atoms on the left side and only one oxygen atom on the right side. We can balance the oxygen atoms by placing a 2 in front of MgO, which gives us two oxygen atoms on each side of the equation.

$Mg + O_2 \longrightarrow 2MgO$ (unbalanced)

Now the magnesium atoms are unbalanced. We balance the magnesium atoms and the whole equation by placing a 2 in front of Mg.

$2\,Mg + O_2 \longrightarrow 2MgO$ (balanced)

The equations in examples 2 and 3 are balanced as follows:

$$BaCl_2 + H_2SO_4 \longrightarrow BaSO_4 + 2HCl$$

$$Na_2CO_3 + 2HCl \longrightarrow 2NaCl + CO_2 + H_2O$$

Remember! A correct formula for a substance may not be changed for the convenience of balancing an equation. Coefficients, as needed, are placed in front of the formulas to balance the equation. However, it is important to notice that when a number is placed in front of one formula to balance a particular element, as in 2 MgO above, it may unbalance another element in the equation (which is why the 2 was added to the Mg)

Note: a coefficient in front of a formula multiplies every atom in the formula by that number. Thus 2 MgO means 2 Mg atoms and 2 O atoms; 3 $CaCl_2$ means 3 Ca atoms and 6 Cl atoms; 4 H_2SO_4 means 8 H atoms, 4 S atoms, and 16 O atoms; 3 $Cu(NO_3)_2$ means 3 Cu atoms, 6 N atoms, and 18 O atoms or 3 Cu atoms and 6 NO_3^- ions.

Many equations may be balanced by this "trial and error" or inspection method.

STUDY AID 3

Preparing and Reading A Graph

A graph is often the most convenient way to present or display a set of data. Various kinds of graphs have been devised, but the most common type uses a set of horizontal and vertical coordinates, x and y, to show the relationship between two variables, the independent and dependent variables. The dependent variable is a measurement that changes as a result of changes in the independent variable. The independent variable either changes itself (like time) or is controlled by the experimenter. Usually the independent variable is plotted on the x-axis (abscissa) and the dependent variable is plotted on the y-axis (ordinate). See Figure S3.1.

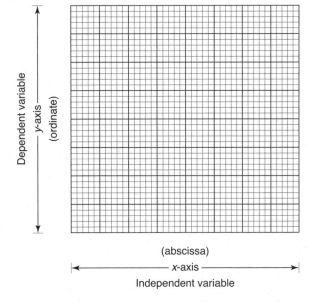

Figure S3.1   Rectangular coordinate graph paper

The values for each variable are called data and listed in a data table to facilitate the construction of a graph. As a specific example of how a graph is constructed, let us graph the relationship between the volume of a liquid and its mass. A chemist measured increasing volumes of a liquid and determined the mass of each volume. The data are recorded in Table S3.1. In this study aid, we will use this data to illustrate the steps for making graphs by hand (Part A) and by computer (Part C).

A. STEPS IN PREPARING A GRAPH

Most scientists today use computers to help them make graphs from their data. Before we show you how to work with a computer to do this, it is important to learn how to make a graph with pencil, ruler, and graph paper. Use the following step-by-step procedure to plot the data in Table S3.1 on the graph paper provided in Figure S3.2. Your completed graph should resemble the graph in Figure S3.3 very closely. After you complete this first graph, practice your graphing skills by making another graph using the data in Table S3.2 and the grid provided in Figure S3.4.

PROCEDURE

1. Examine the graph paper in Figure S3.2 and count how many blocks are available along each axis: This paper has 40 blocks along the x-axis and 40 blocks along the y-axis.

Table S3.1 Volume vs. Mass Data

Volume, mL	Mass, g
21.0	19.1
30.0	27.3
37.5	34.1
44.0	40.0
47.0	42.8
50.0	45.5

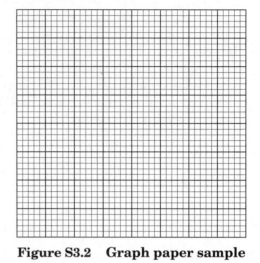

Figure S3.2 Graph paper sample

2. Examine the data in Table S3.1 and determine the independent and dependent variables: The amount of liquid in each sample was varied by the experimenter, so volume is the independent variable and will be plotted along the x-axis. The mass of each sample changed as the volume was changed, so mass is the dependent variable and will be plotted along the y-axis. Usually the independent-versus-dependent-variable decision can be reasoned out like this example. If not, then the placement of the variables on the axes can be arbitrary.

3. Determine the range for each variable: The independent variable ranges from 21.0 mL to 50.0 mL. This is a range of 29.0 mL. The dependent variable ranges from 19.1 g to 45.5 g. This is a range of 26.4 g.

4. Determine the scale for each axis; that is, how many units each block will represent. The calculation for the independent variable using this particular piece of graph paper is:

independent variable scale = 29.0 mL/40 blocks = 0.73 mL/block

But, if we adopted this scale, the graph would be extremely awkward to plot and read. So, we round **up** (never round down) this preliminary scale to a more convenient value per block. The most convenient scales to use are generally 0.5, 1, 2, 5, or 10 units per block. The scale is never rounded up to more than double its preliminary value. For this sample data, 0.73 mL/block is rounded up to 1.0 mL/block because it is convenient and less than 1.46 (double 0.73).

Now, we do the same calculations for the dependent variable on the y-axis.

dependent variable scale = 26.4 g/40 blocks = 0.66 g/block. It is not very convenient to count by units of 0.66, so this value is rounded **up** to 1.0 g per block.

5. Determine the starting values for each coordinate: Although it is common for the axes to be numbered starting with zero at the origin (lower left corner), it is not required and

sometimes it is a poor choice. For instance, in our example, all of the data for the independent variable are greater than 20.0, so from 0–20, there would be no data plotted. Therefore, we start numbering the x-axis at 20.0 mL and the y-axis at 15.0 g.

6. Determine the major and minor increments for each axis: We never number every block. Instead we number in major increments of several blocks with minor, unnumbered increments (blocks) in between. Because of our choice of scales, we will label both the x-axis and the y-axis every 5 blocks. The axes do not have to be numbered every 5 divisions, but often the graph paper has darker lines every five blocks and it is convenient to number at these heavier lines. The numbered increments must be on lines.

7. Label each axis so it is clear what each one represents: In our example, we label the x-axis as Volume, mL, and the y-axis as Mass, g. Labels and units on the coordinates are absolutely essential.

8. Plot the data points: Here is how a point is located on the graph: Using the 44.0 mL and 40.0 g data as an example, trace a vertical line up from 44.0 mL on the x-axis and a horizontal line across from 40.0 g on the y-axis and mark the point where the two lines intersect. This process is called plotting. The remaining five points are plotted on the graph in the same way. It is often helpful if each data point is neatly circled so it will be more visible. Then, if more than one set of data is plotted on the same graph, another symbol (an open triangle or square, for example) can be used.

9. Draw a smooth line through the plotted points: In our example, if the six points have been plotted correctly, they lie on a straight line so that a straight edge can be used to draw the smooth line connecting the six points. When plotting data collected in the laboratory, the best smooth line will not necessarily touch each of the plotted points. Thus, some judgment must be exercised in locating the best smooth line, whether it be straight or curved.

10. Title the graph: Every graph should have a title that clearly expresses what the graph represents. Titles may be placed above the graph or on the upper part of the graph. The latter choice, which is illustrated in Figure S3.3, is the most common for student laboratory reports. Of course, the title must be placed so as not to interfere with the plot on the graph. A completed graph of the data in Table S3.1 is shown in Figure S3.3.

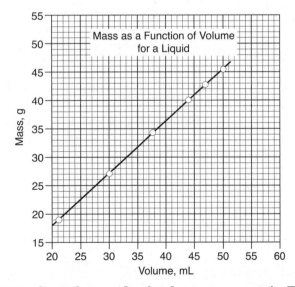

Figure S3.3   Sample graph of volume vs. mass in Table S3.1

Practice Plotting: Sample Data

Table S3.2 is a set of data for you to practice plotting a graph with the steps just described. Use the graph paper in Figure S3.4. Plot °C on the *x*-axis and °F on the *y*-axis.

Table S3.2
Temperature Scales

Temperature, °C	Temperature, °F
0	32
20	68
37	98.6
50	122
100	212

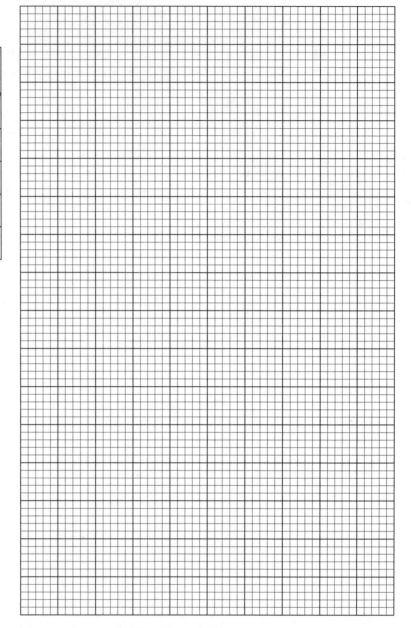

Figure S3.4   Grid for practice plotting of Table S3.2 temperature data

B.   READING A GRAPH

Although graphs are prepared from a limited number of data points (the graph in Figure S3.5 was prepared from six data points), it is possible to extract reliable data for points between the experimental data points and to infer information beyond the range of the plotted data. These skills require that you understand how to read a graph.

Table S3.3

Temperature °C	Solubility, g KClO₃/100 g water
10	5.0
20	7.4
30	10.5
50	19.3
60	24.5
80	38.5

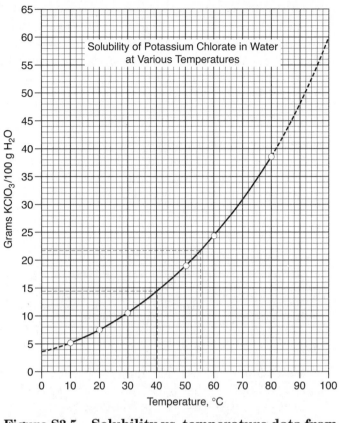

Figure S3.5 Solubility vs. temperature data from Table S3.3

Figure S3.5 is a graph showing the solubility of potassium chlorate in water at various temperatures. The solubility curve on this graph was plotted from experimentally determined solubilities at six temperatures shown in Table S3.3.

These experimentally determined solubilities are all located on the smooth curve traced by the solid line portion of the graph. We are therefore confident that the solid line represents a very good approximation of the solubility data for potassium chlorate covering the temperature range from 10°C to 80°C. All points on the plotted curve represent the composition of saturated solutions. Any point below the curve represents an unsaturated solution.

The dashed line portions of the curve are **extrapolations;** that is, they extend the curve above and below the temperature range actually covered by the plotted data. Curves such as this are often extrapolated a short distance beyond the range of the known data although the extrapolated portions may not be highly accurate. Extrapolation is justified only in the absence of more reliable information.

The graph in Figure S3.5 can be used with confidence to obtain the solubility of $KClO_3$ at any temperature between 10°C and 80°C but the solubilities between 0°C and 10°C and between 80°C and 100°C are less reliable. For example, what is the solubility of $KClO_3$ at 40°C, at 55°C, and at 100°C? First, draw a vertical line from each temperature to the plotted solubility curve. Now from each of these points on the curve, draw a horizontal line to the solubility axis and read the corresponding solubility. The values that we read from the graph are

40°C	14.6 g $KClO_3$/100 g water
55°C	21.9 g $KClO_3$/100 g water
100°C	59.8 g $KClO_3$/100 g water

Of these solubilities, the one corresponding to 55°C is probably the most reliable because experimental points are plotted at 50°C and 60°C. The 40°C solubility value is probably a bit less reliable because the nearest plotted points are at 30°C and 50°C. The 100°C solubility is the least reliable of the three values because it was taken from the extrapolated part of the curve, and the nearest plotted point is 80°C. Actual handbook solubility values are 14.0 g and 57.0 g of $KClO_3$/100 g water at 40°C and 100°C respectively.

Although making and reading graphs by hand in this way gets easier with practice, most scientists now use computers to graph their experimental data. If you have access to a computer either in the laboratory, library, or at home, we encourage you to learn computer graphing after you have mastered all the skills of graphing data by hand. Instructions for computer graphing are provided in Part C of this study aid.

Requirements for Computer Graphing

There are several software programs that can be used to generate graphs of scientific data. We have chosen **Microsoft Excel,** a program within *Microsoft Office 2007* (earlier versions will also work) which requires PC Windows XP or Vista. You should be sitting in front of a PC computer with Microsoft Excel 2007 active as you begin these instructions. These instructions assume that you have some knowledge of Microsoft Excel and that you know how to write graphs manually as instructed in Part A of this Study Aid. Unless you are comfortable with your computer and have used this software, there should be someone around who can help you out occasionally when you try this the first time.

C. COMPUTER GRAPHING

Preparing a computer graph involves the same steps as the paper and pencil graph. The difference is that the computer will do most of the steps for you. However, you must provide the computer with the necessary information (data) that is being graphed. After you complete all the instructions in this section, you should have a graph that looks like Figure **S3.6** below.

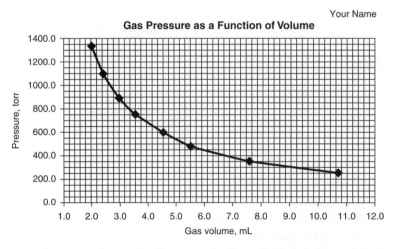

Figure S3.6   Sample Computer Graph Using Excel 2007

1. Open up a clean page (called an Excel "sheet") on which to begin the graph (called a "chart")

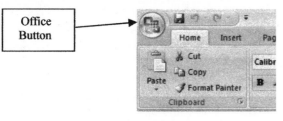

Figure S3.7

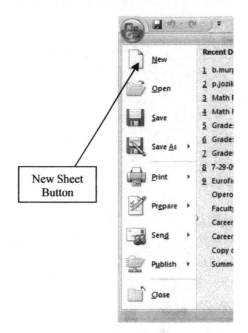

To do this, click on the office button in the upper left corner of the Excel ribbon (**Figure S3.7**). When you click on the Office Button you will open a list of functions (Figure S3.8). Click on the **New** icon and an Excell sheet as shown in Figure S3.9 will open.

Figure S3.8

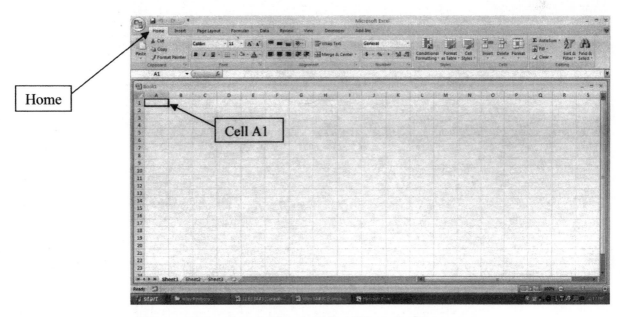

Figure S3.9   Excel Sheet with ribbon in Home open

2. Add the data table to the sheet so the graphing of the data can begin. The data graphed for this exercise is the same data that was graphed in Exercise 9, B2.

Volume, mL	10.70	7.64	5.57	4.56	3.52	2.97	2.43	2.01
Pressure, torr	250	350	480	600	760	900	1100	1330

Go to Column A on the new Excel sheet that was opened. In cell A1, type **Volume, mL** and in B1, type **Pressure, torr.** Then, enter the volume and pressure values in the columns from the data shown above. Highlight the cells that are filled in with the data. When this is done, the left side of your screen should look like **Figure S3.10**.

3. Convert the data table into a "chart" by choosing **insert** from the ribbon (**Figure S3.11**). This should immediately result in the appearance of the tool bar shown in S3.11. Click on Scatter and the menu of charts derived from this category will appear as shown below. For this set of volume and pressure data which is made up of separate measurements, select  and a graph of the volume vs. pressure data in Figure S3.10 will appear on the sheet (Figure S3.12) below.

Anytime you want to modify the chart, you must select the chart by clicking on a blank area of the chart. The border around the graph will appear (as shown in Figure S3.12). This will activate the Chart ribbon with the tools to modify the chart.

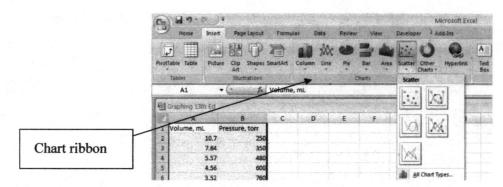

Figure S3.10   Data to be graphed

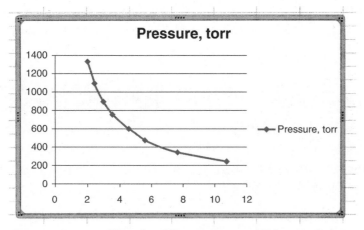

Figure S3.11   Chart ribbon

Figure S3.12   Graph of Pressure vs. Volume data before formatting

4. **Adding a Chart Title and Axis Titles.** The remaining steps in this procedure will explain how to convert this very primitive graph into the finished graph at the beginning

of the Computer Graphing instructions (**Figure S3.6**). Make sure the Chart ribbon is open, and select Chart Tools as shown below.

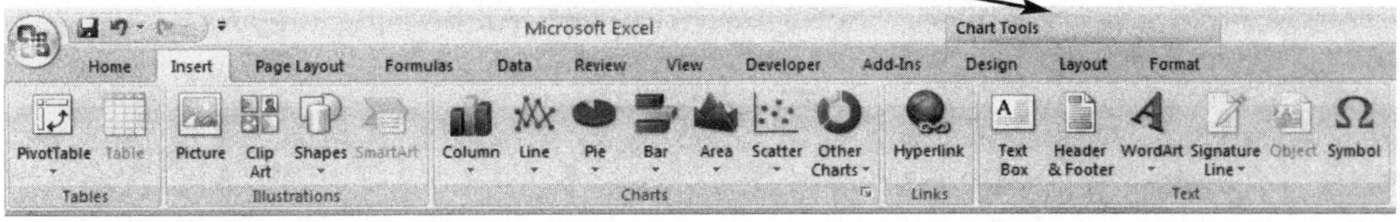

Select Layout from the Chart Tools options and a chart layout ribbon will appear as shown in Figure S3.13

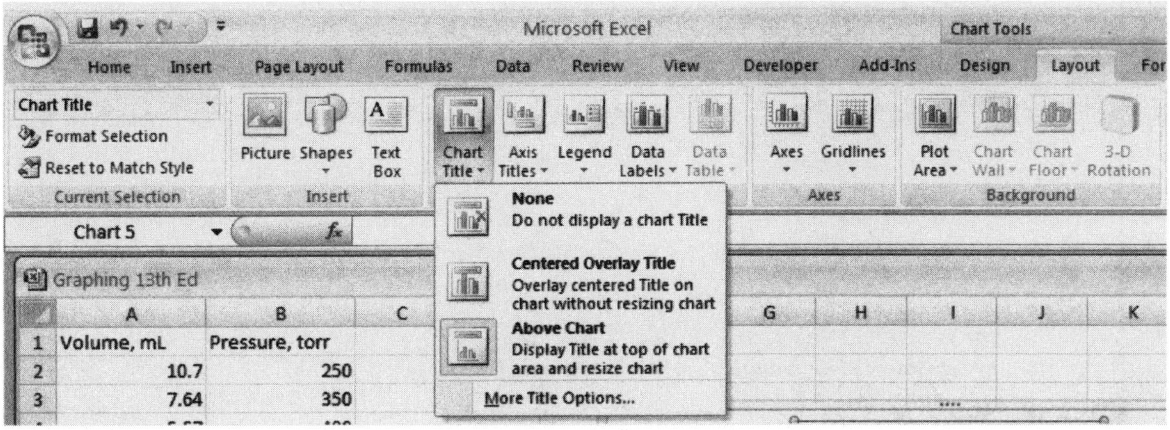

Figure S3.13

The Chart ribbon shown above will be used to complete the formatting of the graph. Choose Chart Title and the menu will appear as shown in S3.13. Click on Above Chart as shown and a text box will appear on the chart. Fill in an appropriate title for the chart: **Gas Pressure as a Function of Volume**.

5. **Adding Titles for the X and Y axes**. Select Axis Titles from the Chart Layout ribbon. From the drop down menus, select Primary Horizontal Axis Title (ie. X axis) and Title Below Axis.

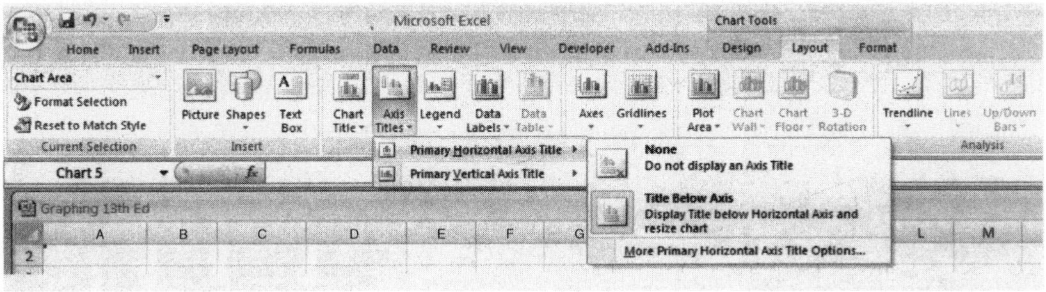

Figure S3.14

6. This will bring up a text box for adding the X-axis title: **Gas volume, mL**. Next, select Primary vertical Axis title and select **Rotated title**: rotated axis title and resize chart. This will open up a text box for adding the Y-axis title: **Pressure, torr**.

7. **Format Legend.** A legend is a table which shows the symbol and line corresponding to each set of data points (**Figure S3.14**). Since there is only one kind of symbol and one line in this graph, the legend is not needed. Therefore, select ⎡Legend⎤ and choose ⎡None⎤ to remove the default legend. If you choose to keep the legend, you can select where it will be placed on the chart from this same menu.

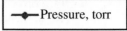

Figure S3.15

8. **Format Axis Scale, decimal places, and other options.** Select ⎡Axes⎤ from the chart layout ribbon and a menu will appear as shown in **Figure S3.16.** Select ⎡Primary horizontal axis⎤ and from that menu select ⎡More Primary horizontal axis options⎤. A menu as shown in **Figure S3.16** should appear. For the data in this exercise, select the options shown on the right.

Figure S3.16

Format the y axis scale the same way. Go back to Axis on the Chart Layout ribbon. ⎡Select Primary vertical axis⎤ and then select ⎡More primary vertical axis options⎤. You will get a menu similar to Fig. 3.14. Complete the Y axis options as follows: Minimum = fixed, 0; Maximum = fixed, 1400; Major unit = fixed, 200; Minor unit = fixed, 50. Major tick mark type: Outside; Minor tick mark type: Inside; Axis labels: Next to axis. Vertical axis crosses: Automatic.

9. **Format the number type:** Select ⎡Number⎤ from the Format Axis menu. From the list of choices select ⎡Number⎤, decimal places 1, and 1000 separater, ⎡none⎤ (you will need to deselect the √ for this (which means there is a comma in the 1000 place (ex. 1,000).

10. Other format axis options: these can be left in the default positions. No need to change them.

11. Add gridlines which make the graph appear like it was made on graph paper. Select Gridlines from the Chart Layout ribbon (**Figure S3.17**). For both Primary horizontal and vertical gridlines, select ⎡Major and Minor Gridlines⎤.

Figure S3.17

12. **Change the color and border of the plot area**. Select [Plot Area] from the Chart Layout ribbon to make changes to the fill inside the plot area or to change the border around the plot area. Select More Plot Area Options to vary the Fill, Border, and Format of the plot area. The sample graph is in the default fill color and border.

13. **More options.** There are many more Excel features that can be used to customize a graph of data. You are encouraged to explore these options as you use computer graphing for this course. For example, some data is best graphed with a **line of best fit**. This can be done by selecting Scatter (without a line) from the Chart ribbon and then Trendline from the Chart Layout ribbon.

14. **Add your name to the graph**. Select the graph by clicking on a blank space. Click on the Insert tab then click on [Header] [& Footer]. The Page set-up menu will be displayed (**Figure S3.18**). Select Custom Header and the Header Box will appear. Type your name and any other information the instructor has requested in the [Right section] of the Header box and click OK (twice).

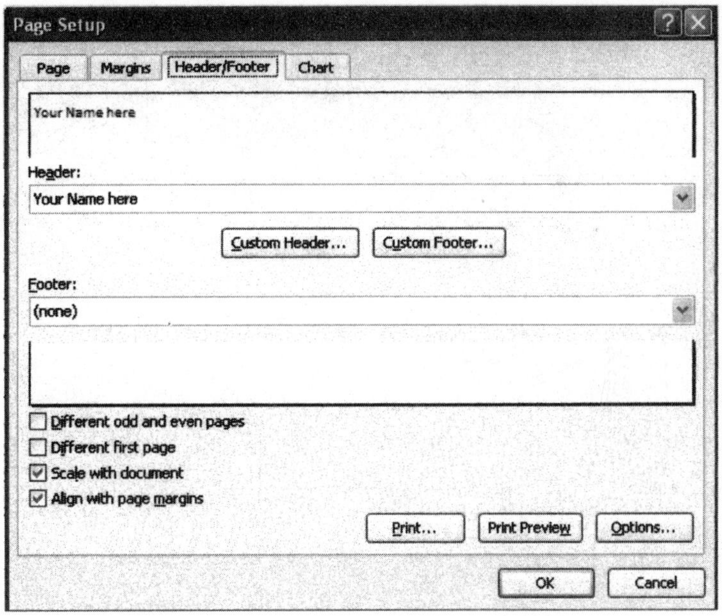

15. **Print your graph.** Select the Office icon on the top left corner of the screen. Select Print from the menu and your graph will be printed. Do not close your computer graphing work until your graph is printed in as many copies as you need.

Figure S3.18

STUDY AID 4

Using a Scientific Calculator

A calculator is useful for most calculations in this book. You should obtain a scientific calculator; that is, one that has at least the following function keys on its keyboard.

Addition $\boxed{+}$

Subtraction $\boxed{-}$

Multiplication $\boxed{\times}$

Division $\boxed{\div}$

Equals $\boxed{=}$

Second function $\boxed{\text{2nd}}$ or $\boxed{\text{INV}}$ or $\boxed{\text{Shift}}$

Change sign $\boxed{+/-}$

Exponential number $\boxed{\text{Exp}}$

Logarithm $\boxed{\text{Log}}$

Antilogarithm $\boxed{10^{\times}}$

Mode $\boxed{\text{MODE}}$

All calculators do not use the same symbolism for these function keys. Not all calculators work the same way. Save the instruction manual that comes with your calculator. It is very useful for determining how to do special operations on your particular model. Refer to your instruction manual for variations from the function symbols shown above and for the use of other function keys.

Some keys have two functions, upper and lower. In order to use the upper (second) function, the second function key $\boxed{\text{2nd}}$ must be pressed in order to activate the desired upper function.

The display area of the calculator shows the numbers entered and often shows more digits in the answer than should be used. The numbers in the display can be in fixed decimal form or in exponential notation, depending on how the calculator is programmed. The MODE key on many calculators is used to change back and forth between fixed decimal and exponential notation. Refer to your instruction manual for how to use this function. Regardless of the digits in the display, the final answer should always be rounded to reflect the proper number of significant figures for the calculations. The calculator will not do that for you.

Addition and Subtraction

To add numbers:

1. Enter the first number to be added followed by the plus key $\boxed{+}$.

2. Enter the second number to be added followed by the plus key $\boxed{+}$.

3. Repeat Step 2 for each additional number to be added, except the last number.

4. After the last number is entered, press the equal key $\boxed{=}$. You should now have the answer in the display area.

5. When a number is to be subtracted, use the minus key $\boxed{-}$ instead of the plus key.

As an example, to calculate 16.0 + 1.223+8.45, enter 16.0 followed by the ⊞ key; then enter 1.223 followed by the ⊞ key; then enter 8.45 followed by the ⊟ key. The display shows 25.673, which is rounded to the answer 25.7.

Examples of Addition and Subtraction

Calculation	Enter in Sequence	Display	Rounded Answer
a. 12.0 + 16.2 + 122.3	12.0 ⊞ 16.2 ⊞ 122.3 ⊟	150.5	150.5
b. 132 − 62 + 141	132 ⊟ 62 ⊞ 141 ⊟	211	211
c. 46.23 + 13.2	46.23 ⊞ 13.2 ⊟	59.43	59.4
d. 129.06 + 149.1 − 18.3	129.06 ⊞ 49.1 ⊟ 18.3 ⊟	159.86	159.9

Multiplication

To multiply numbers using your calculator

1. Enter the first number to be multiplied followed by the multiplication key ⊠ .

2. Enter the second number to be multiplied followed by the multiplication key ⊠ .

3. Repeat Step 2 for all other numbers to be multiplied except the last number.

4. Enter the last number to be multiplied followed by the equal key ⊟ . You now have the answer in the display area. Round off to the proper number of significant figures. As an example, to calculate (3.25)(4.184)(22.2) enter 3.25 followed by the ⊠ key; then enter 4.184 followed by the ⊠ key; then enter 22.2 followed by the ⊟ key. The display shows 301.8756, which is rounded to the answer 302.

Examples of Multiplication

Calculation	Enter in Sequence	Display	Rounded Answer
a. (12)(14)(18)	12 ⊠ 14 ⊠ 18 ⊟	3024	3.0×10^3
b. (122)(3.4)(60.)	122 ⊠ 3.4 ⊠ 60. ⊟	24888	2.5×10^4
c. (0.522)(49.4)(6.33)	0.522 ⊠ 49.4 ⊠ 6.33 ⊟	163.23044	163

Division

To divide numbers using your calculator:

1. Enter the numerator followed by the division key ÷ .

2. Enter the denominator followed by the equal key $\boxed{=}$ to give the answer in the display area. Round off to the proper number of significant figures.

3. If there is more than one denominator, enter each denominator followed by the division key except for the last number, which is followed by the equal key. As an example, to calculate $\dfrac{126}{12}$, enter 126 followed by the $\boxed{\div}$ key; then enter 12 followed by the $\boxed{=}$ key. The display shows 10.5, which is rounded to the answer 11.

Examples of Division

	Calculation	Enter in Sequence	Display	Rounded Answer
a.	$\dfrac{142}{25}$	142 $\boxed{\div}$ 25 $\boxed{=}$	5.68	5.7
b.	$\dfrac{0.422}{5.00}$	0.422 $\boxed{\div}$ 5.00 $\boxed{=}$	0.0844	0.0844
c.	$\dfrac{124}{(0.022)(3.00)}$	124 $\boxed{\div}$ 0.022 $\boxed{\div}$ 3.00 $\boxed{=}$	1878.7878	1.9×10^3

Exponents

In scientific measurements and calculations we often encounter very large and very small numbers. A convenient method of expressing these large and small numbers is by using exponents or powers of 10. A number in exponential form is treated like any other number, that is, it can be added, subtracted, multiplied, or divided.

To enter an exponential number into your calculator first enter the non-exponential part of the number, then press the exponent key $\boxed{\text{Exp}}$, followed by the exponent. For example, to enter 4.94×10^3, enter 4.94, then press $\boxed{\text{Exp}}$, then press 3. When the exponent of 10 is a minus number, press the Change of Sign key $\boxed{+/-}$ after entering the exponent. For example, to enter 4.94×10^{-3}, enter in sequence 4.94 $\boxed{\text{Exp}}$ 3 $\boxed{+/-}$. In most calculators the exponent will appear in the display a couple of spaces after the non-exponent part of the number; for example, 4.94   03 or 4.94   −03.

Examples Using Exponential Numbers

	Calculation	Enter in Sequence	Display	Rounded Answer
a.	$(4.94 \times 10^3)(21.4)$	4.94 $\boxed{\text{Exp}}$ 3 $\boxed{\times}$ 21.4 $\boxed{=}$	105716	1.06×10^5
b.	$(1.42 \times 10^4)(2.88 \times 10^{-5})$	1.42 $\boxed{\text{Exp}}$ 4 $\boxed{\times}$ 2.88 $\boxed{\text{Exp}}$ 5 $\boxed{+/-}$ $\boxed{=}$	0.40896	0.409
c.	$\dfrac{8.22 \times 10^{-5}}{5.00 \times 10^7}$	8.22 $\boxed{\text{Exp}}$ 5 $\boxed{+/-}$ $\boxed{\div}$ 5.00 $\boxed{\text{Exp}}$ 7 $\boxed{=}$	1.644 −12	1.64×10^{-12}

Logarithms

The logarithm (log) of a number to the base 10 is the power (exponent) to which 10 must be raised to give that number. For example, the log of 100 is 2.0 (log $100 = 10^{2.0}$). The log of 200 is 2.3 (log $200 = 10^{2.3}$). Logarithms are used in chemistry to calculate the pH of an aqueous solution. The answer (log) should contain the same number of significant figures to the right of the decimal as there are significant figures in the original number. Thus the log $100 = 2.0$ but the log 100. is 2.000.

The log key on many calculators is a second function key. To determine the log using your calculator, enter the number, then press the log key $\boxed{\text{Log}}$. For example to determine the log of 125, enter 125, then press the log key $\boxed{\text{Log}}$. The display shows 2.09691, which is rounded to the answer 2.097.

Examples. Determine the log of the following:

Calculation	Enter in Sequence	Display	Rounded Answer
a. log 42	42 $\boxed{\text{Log}}$	1.6232492	1.62
b. log 1.62×10^5	1.62 $\boxed{\text{Exp}}$ 5 $\boxed{\text{Log}}$	5.209515	5.210
c. log 6.4×10^{-6}	6.4 $\boxed{\text{Exp}}$ 6 $\boxed{+/-}$ $\boxed{\text{Log}}$	−5.19382	−5.19
d. log 2.5	2.5 $\boxed{\text{Log}}$	0.39794	0.40

Antilogarithms (Inverse Logarithms)

An antilogarithm is the number from which the logarithm has been calculated. It is calculated using the to $\boxed{10^\times}$ key on your calculator. Many calculators use the $\boxed{\text{2nd}}$ or $\boxed{\text{INV}}$ or $\boxed{\text{Shift}}$ key to access this function. To use a function that appears above a key you must first press the $\boxed{\text{2nd}}$ key. For example, to determine the antilogarithm of 2.891, enter 2.891 into your calculator, then press the $\boxed{\text{2nd}}$ key followed by the $\boxed{10^\times}$ key. The display shows 778.03655, which is rounded to the answer 778. The answer should contain the same number of figures as there are to the right of the decimal in the antilog.

Examples. Determine the antilogarithm of the following:

Calculation	Enter in Sequence	Display	Rounded Answer
a. antilog 1.628	1.628 $\boxed{\text{2nd}}$ $\boxed{10^\times}$	42.461956	42.5
b. antilog 7.086	7.086 $\boxed{\text{2nd}}$ $\boxed{10^\times}$	12189896	1.22×10^7
c. antilog −6.33	6.33 $\boxed{+/-}$ $\boxed{\text{2nd}}$ $\boxed{10^\times}$	4.6773514 −07	4.7×10^{-7}

Additional Practice Problems

Only the problem, the display, and the answers are given.

Problem	Display	Answer
1. $143.5 + 14.02 + 1.202$	158.722	158.7
2. $72.06 - 26.92 - 49.66$	-4.52	-4.52
3. $2.168 + 4.288 - 1.62$	4.836	4.84
4. $(12.3)(22.8)(1.235)$	346.3434	346
5. $(2.42 \times 10^6)(6.08 \times 10^{-4})(0.623)$	916.65728	917
6. $\dfrac{(46.0)(82.3)}{19.2}$	197.17708	197
7. $\dfrac{0.0298}{243}$	1.2263374 $\quad -04$	1.23×10^{-4}
8. $\dfrac{(5.4)(298)(760)}{(273)(1042)}$	4.2992554	4.3
9. $(6.22 \times 10^6)(1.45 \times 10^3)(9.00)$	8.1171 $\quad$ 10	8.12×10^{10}
10. $\dfrac{(1.49 \times 10^6)(1.88 \times 10^6)}{6.02 \times 10^{23}}$	4.6531561 $\quad -12$	4.65×10^{-12}
11. $\log 245$	2.389166	2.389
12. $\log 6.5 \times 10^{-6}$	-5.1870866	-5.19
13. $(\log 24)(\log 34)$	2.1137644	2.11
14. antilog 6.34	2187761.6	2.2×10^6
15. antilog -6.34	4.5708819 $\quad -07$	4.6×10^{-7}

STUDY AID 5

Dimensional Analysis and Stoichiometry

Chemistry is a quantitative science and involves measurements, complex calculations, and problem solving. Dimensional analysis is an important tool of chemistry, just like the electronic calculator, and measuring devices (like the thermometer, buret, and graduated cylinder).

In dimensional analysis the units for all quantities are always carried along with their corresponding number, the units for the answer come out of the calculations automatically, and errors in the reasoning behind a series of calculations are easily identified and corrected.

Dimensional analysis as a problem-solving tool has many applications in chemistry and is used in many of the laboratory experiments in this manual including: the conversion of one unit into another, calorimetry, solution concentrations, moles and stoichiometry, gas laws, and heat of reaction. Regardless of the application, the basis of dimensional analysis is the use of conversion factors to organize a series of steps in quest of a specific quantity with a specific unit.

A.   Conversion Factors

Conversion factors come from equivalent relationships or ratios between two quantities. These relationships are usually expressed as equations or derived units. When used as conversion factors, they are written in fractional form. Some specific examples are shown below:

Example Equivalence Statement or Derived Unit	Conversion Factor #1	Conversion Factor #2
1 mole H_2O = 1 molar mass H_2O	$\dfrac{1 \text{ mole } H_2O}{18.01 \text{ g } H_2O}$	$\dfrac{18.01 \text{ g } H_2O}{1 \text{ mole } H_2O}$
1 atmosphere = 760 mm Hg or 760 torr	$\dfrac{1 \text{ atm}}{760 \text{ mm Hg}}$	$\dfrac{760 \text{ mm Hg}}{1 \text{ atm}}$
4.184 J/g°C (Specific heat of water)	$\dfrac{4.184 \text{ J}}{\text{g°C}}$	$\dfrac{1 \text{ g°C}}{4.184 \text{ J}}$
8.96 g Cu/1 mL (density of Cu)	$\dfrac{8.96 \text{ g Cu}}{1 \text{ mL}}$	$\dfrac{1 \text{ mL}}{8.96 \text{ g Cu}}$
22.4 L = 1 mol gas at STP	$\dfrac{22.4 \text{ L}}{1 \text{ mol gas}}$	$\dfrac{1 \text{ mol gas}}{22.4 \text{ L}}$

B.   Unit Conversions

The dimensional analysis method of converting units involves organizing one or more conversion factors into a logical series that cancels or eliminates all units except the unit(s) wanted in the answer.

The calculation setup requires two conversion factors: lb $\longrightarrow$ g $\longrightarrow$ mL

$$(0.15\,\text{lb Cu})\left(\frac{453.6\,\text{g}}{1\,\text{lb}}\right)\left(\frac{1\,\text{mL Cu}}{8.96\,\text{g Cu}}\right) = 7.6\,\text{mL Cu}$$

Note, that in completing this calculation, units are treated as numbers, lb in the denominator are canceled into lb in the numerator and g in the denominator are canceled into g in the numerator.

Problem 2. How many grams of sodium chloride are in 0.250 L of a solution with a density of 1.04 g/mL that is 10.0% sodium chloride?

The calculation setup requires three conversion factors:

$$\text{L} \longrightarrow \text{mL} \longrightarrow \text{g NaCl}(aq) \longrightarrow \text{g NaCl}$$

$$(0.250\,\text{L NaCl}(aq))\left(\frac{1000\,\text{mL}}{1\,\text{L}}\right)\left(\frac{1.04\,\text{g}}{1\,\text{mL}}\right)\left(\frac{10.0\,\text{g NaCl}}{100.0\,\text{g NaCl}(aq)}\right) = 26.0\,\text{g NaCl}$$

C.   The Mole and Stoichiometry

It is often necessary to calculate the amount of product that can be obtained from a given amount of reactant or, conversely, to determine how much reactant is required to produce a stated amount of product. Calculations of this kind, based on balanced chemical equations, are called **stoichiometry** (from Greek, meaning element measure).

In solving stoichiometric problems, conversion factors based on **the mole** are very important. In its broadest sense a mole is Avogadro's number (6.022×10^{23}) of any chemical species. Even though the unit "mole" is used as a short expression for molar mass, it is quite permissible to refer to moles of chemical species that are not really molecular in character. Reference may be made to moles of such diverse species as sulfur atoms (S), oxygen atoms (O), oxygen molecules (O_2), sulfuric acid molecules (H_2SO_4), sodium chloride formula units (NaCl), ammonium ions (NH_4^+), nitrate ions (NO_3^-), or even to moles of electrons or protons.

From these definitions of mole we can write two equivalence statements on which to base conversion factors.

1 mole $= 6.022 \times 10^{23}$ items

1 mole = molar mass in g (mass in grams numerically equal to the molar mass)

Three more conversion factors based on the mole are also useful. One applies only to gases: 1 mole = 22.4 L of gas at STP. The rationale for this is based on Avogadro's famous law that "equal volumes of all gases, at the same temperature and pressure, contain the same number of molecules". The second applies to solutions for which the concentration is expressed as **molarity**. Molarity is defined as the number of moles of solute in 1 L of solution. The last is

based on the mole ratios of reactants and products in a balanced equation. For example, in the hypothetical reaction

$$2A + 3B \longrightarrow A_2B_3$$

the mole ratios of the reactants and product to each other are 2 mol A to 3 mol B to 1 mol A_2B_3. The following table includes the basic equivalence statements or derived units which include the mole. The abbreviation for mole is mol both sigular and plural.

Equivalence Statement	Conversion Factor #1	Conversion Factor #2
1 mol = 6.022×10^{23} items	$\dfrac{1\,mol}{6.022 \times 10^{23}\,items}$	$\dfrac{6.022 \times 10^{23}\,items}{1\,mol}$
1 mol = molar mass in g	$\dfrac{1\,mol}{molar\,mass\,in\,g}$	$\dfrac{molar\,mass\,in\,g}{1\,mol}$
1 mol = 22.4 L of gas at STP	$\dfrac{1\,mol}{22.4\,L\,at\,STP}$	$\dfrac{22.4\,L\,at\,STP}{1\,mol}$
mol solute/1 L solution	$\dfrac{mol\,solute}{1\,L\,of\,solution}$	$\dfrac{1\,L\,of\,solution}{mol\,solute}$
mol species A/mol species B	$\dfrac{mol\,A}{mol\,B}$	$\dfrac{mol\,B}{mol\,A}$

Using dimensional analysis involving mole conversion factors to solve a problem requires four steps:

1. Write the balanced chemical equation for the reaction involved.

2. Examine the problem statement and determine what is the given substance that will be the starting point for the calculation.

3. Set up a series of conversion factors that eliminate (by cancellation) all units except the unit specified for the answer.

4. Do the calculations and express the answer with the correct number of significant figures.

Apply these steps to the following problems related to the chemical decomposition of potassium chlorate to produce potassium chloride and oxygen.

Problem 3. How many grams of oxygen can be obtained from 15.0 g of $KClO_3$?

1. Balanced equation: $2\,KClO_3 \longrightarrow 2\,KCl + 3\,O_2$

2. The problem states that 15.0 g of $KClO_3$ is being converted to O_2. Therefore, the dimensional analysis setup begins with 15.0 g $KClO_3$

3. The calculation setup will be determined by choosing conversion factors that cancel units in the preceding quantity or conversion factor. For this problem, three conversion factors are needed:

$$g\,KClO_3 \longrightarrow mol\,KClO_3 \longrightarrow mol\,O_2 \longrightarrow g\,O_2$$

$$(15.0\,\cancel{g\,KClO_3})\left(\frac{1\,\cancel{mol\,KClO_3}}{122.6\,\cancel{g\,KClO_3}}\right)\left(\frac{3\,\cancel{mol\,O_2}}{2\,\cancel{mol\,KClO_3}}\right)\left(\frac{32.00\,g\,O_2}{1\,\cancel{mol\,O_2}}\right)$$

(Note: 122.6 g and 32.00 g are the molar masses of $KClO_3$ and O_2, respectively).

4. Once all the units have been canceled except the units specified for the answer, the calculations can be completed.

$$\frac{(15.0)(3)(32.00)\,g\,O_2}{(122.6)(2)} = \textbf{5.87 g O}_\textbf{2}$$

Problem 4. How many oxygen molecules can be obtained from 15.0 g of $KClO_3$?

Steps 1,2 and 4 are exactly the same as in Problem 3, but the last conversion in step 3 requires a different conversion factor:

$$g\,KClO_3 \longrightarrow mol\,KClO_3 \longrightarrow mol\,O_2 \longrightarrow molecules\,O_2$$

$$(15.0\,\cancel{g\,KClO_3})\left(\frac{1\,\cancel{mol\,KClO_3}}{122.6\,\cancel{g\,KClO_3}}\right)\left(\frac{3\,\cancel{mol\,O_2}}{2\,\cancel{mol\,KClO_3}}\right)\left(\frac{6.022 \times 10^{23}\,molecules\,O_2}{1\,\cancel{mol\,O_2}}\right)$$

$$= \textbf{1.11} \times \textbf{10}^{\textbf{23}}\ \textbf{molecules O}_\textbf{2}$$

Problem 5. How many liters of oxygen gas, measured at STP, can be obtained from 15.0 g of $KClO_3$?

Steps 1, 2, and 4 are exactly the same as in Problems 3 and 4, but the last conversion in step 3 requires a different conversion factor:

$$g\,KClO_3 \longrightarrow mol\,KClO_3 \longrightarrow mol\,O_2 \longrightarrow L\,O_2$$

$$(15.0\,\cancel{g\,KClO_3})\left(\frac{1\,\cancel{mol\,KClO_3}}{122.6\,\cancel{g\,KClO_3}}\right)\left(\frac{3\,\cancel{mol\,O_2}}{2\,\cancel{mol\,KClO_3}}\right)\left(\frac{22.4\,L}{1\,\cancel{mol\,O_2}}\right) = \textbf{4.11 L}$$

Problem 6. How many grams of $KClO_3$ must be decomposed to produce 25.0 g of KCl?

Step 1. Balance the equation for the reaction as shown in Problem 3.

Step 2. Consider KCl to be the given substance; convert to g $KClO_3$ using a sequence of conversion factors.

$$g\,KCl \longrightarrow mol\,KCl \longrightarrow mol\,KClO_3 \longrightarrow g\,KClO_3$$

Steps 3 and 4. Choose conversion factors to cancel units of the preceding fractions in the sequence.

$$(25.0\,\text{g}\,\overline{\text{KCl}})\left(\frac{1\,\text{mol}\,\overline{\text{KCl}}}{74.55\,\text{g}\,\overline{\text{KCl}}}\right)\left(\frac{2\,\text{mol}\,\overline{\text{KClO}_3}}{2\,\text{mol}\,\overline{\text{KCl}}}\right)\left(\frac{122.6\,\text{g}\,\text{KClO}_3}{1\,\text{mol}\,\overline{\text{KClO}_3}}\right) = \mathbf{41.1\,g\,KClO_3}$$

Problem 7.   How many mL of 6.0 M HCl(aq) are needed to react with 4.85 g of NaHCO$_3$?

Step 1.   The balanced equation for this reaction is

$$NaHCO_3 + HCl \longrightarrow NaCl + H_2O + CO_2$$

Step 2.   Consider NaHCO$_3$ to be the given substance; convert to mL of 6.0 M HCl using a sequence of conversion factors.

$$g\,NaHCO_3 \longrightarrow mol\,NaHCO_3 \longrightarrow mol\,HCl \longrightarrow L\,HCl \longrightarrow mL\,HCl$$

Steps 3 and 4.   Choose conversion factors to cancel units of the preceding factors in the sequence.

$$(4.85\,\text{g}\,\overline{\text{NaHCO}_3})\left(\frac{1\,\text{mol}\,\overline{\text{NaHCO}_3}}{84.01\,\text{g}\,\overline{\text{NaHCO}_3}}\right)\left(\frac{1\,\text{mol}\,\overline{\text{HCl}}}{1\,\text{mol}\,\overline{\text{NaHCO}_3}}\right)\left(\frac{1\,\overline{\text{L}}}{6.0\,\text{mol}\,\overline{\text{HCl}}}\right)\left(\frac{1000\,\text{mL}}{1\,\overline{\text{L}}}\right)$$

$$= \mathbf{9.6\,mL\,HCl}$$

STUDY AID 6

Organic Chemistry — An Introduction

Organic chemistry is known as the chemistry of the carbon compounds. All compounds that are classified as organic contain carbon. Organic compounds are found in all living matter, food stuffs (fats, proteins, and carbohydrates), fuels of all kinds, plastics, fabrics, wood and paper products, paints and varnishes, dyes, soaps and detergents, cosmetics, medicinals, insecticides, refrigerants, etc. There are over fifty million known organic compounds.

To help study their properties, organic compounds are grouped into classes or series according to the similarity of their chemical makeup or structure. Some of the common classes are hydrocarbons, alcohols, aldehydes, ketones, carboxylic acids, esters, ethers, amines, and amides.

The major reason for the large number of organic compounds is that carbon atoms have the ability to bond together, forming long chains and rings. Carbon atoms share electrons, forming covalent bonds. Between two carbon atoms, single, double, or triple covalent bonds may be formed by sharing one, two, or three pairs of electrons, respectively. These types of bonds are illustrated below:

C:C	C::C	C:::C
C — C	C = C	C ≡ C
Single bond	Double bond	Triple bond

A dash between carbon atoms indicates a covalent bond and represents one pair of electrons.

The names of the alkane series of hydrocarbons are important to organic chemistry because they represent the basis for the systematic nomenclature of organic compounds. The first 10 members of this series and their molecular formulas are listed below:

CH_4 Methane	C_4H_{10} Butane	C_6H_{14} Hexane	C_8H_{18} Octane
C_2H_6 Ethane	C_5H_{12} Pentane	C_7H_{16} Heptane	C_9H_{20} Nonane
C_3H_8 Propane			$C_{10}H_{22}$ Decane

Structural Formulas

A great many organic compounds are composed of carbon, hydrogen, and oxygen atoms. In these compounds we find these atoms bonded to each other in the following ways:

Carbon to carbon	(C — C), (C = C), (C ≡ C)
Carbon to carbon to carbon	(C — C — C)
Carbon to hydrogen	(C — H)
Carbon to oxygen	(C = O)
Carbon to oxygen to hydrogen	(C — O — H)
Carbon to oxygen to carbon	(C — O — C)

In organic compounds, with some exceptions, a carbon atom will have four covalent bonds; a hydrogen atom, one bond; and an oxygen atom, two bonds. A covalent bond consists of a pair of electrons shared between any two atoms.

Because of the different arrangements in which carbon atoms bond with each other, a single written molecular formula may present more than one arrangement of the atoms, giving rise to more than one compound. For example, there are 2 different butanes (C_4H_{10}), 3 pentanes (C_5H_{12}), and 75 decanes ($C_{10}H_{22}$). To illustrate different compounds with the same molecular formula, we use structural formulas. Structural formulas show the order in which the atoms are bonded to each other, while molecular formulas show only the number and kind of each atom in a molecule. A few examples will illustrate.

Methane, CH_4

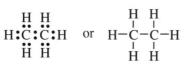

| Structural formula | Condensed structural formula |

Ethane, C_2H_6

| Structural formula | Condensed structural formula |

Note that in the condensed structural formula, which is a convenient simplification of the structural formula, all of the atoms or groups attached to each carbon atom are generally written to the right of it.

Propane, C_3H_8

$$CH_3CH_2CH_4$$

Structural formula   Condensed structural formula

Methyl alcohol, CH_3OH

$$CH_3OH$$

Structural formula   Condensed structural formula

Butane, C_4H_{10} (two isomers)

(a) n-Butane (n = normal)

$$CH_3CH_2CH_2CH_3$$

Structural formula   Condensed structural formula

(b) Isobutane (2-methylpropane)

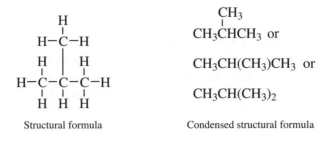

Structural formula   Condensed structural formula

Isomerism

We have shown that there are two possible ways to bond 4 carbon atoms and 10 hydrogen atoms to form the structures of the two butanes. These butanes are indeed different compounds, each with its own physical and chemical properties. For example, n-butane boils at −0.5°C and isobutane boils at −11.7°C.

The phenomenon of two or more compounds having the same molecular formula but different structural formulas is known as **isomerism**. The individual compounds are called **isomers**. Thus there are 2 isomers of butane, 3 of pentane, 18 of octane, and 75 of decane.

Alkyl Groups

The nomenclature of organic chemistry is sprinkled with such terms as methyl, ethyl, and isopropyl. These terms represent **alkyl groups** derived from alkane hydrocarbons. An alkyl group is formed by removing one hydrogen atom from an alkane. For example, the methyl group, CH_3—, is formed by removing one hydrogen from methane, CH_4. The name methyl is formed by dropping the *ane* from the name *methane* and adding the letters *yl* to the remaining stem *meth*. Other alkyl groups are derived in a similar manner—ethyl from ethane, etc. A few of the more common alkyl groups are listed below. The dash in the formula indicates the carbon atom from which the hydrogen atom has been removed.

CH_3—   methyl

CH_3CH_2—   ethyl

$CH_3CH_2CH_2$—   n-propyl (n = normal); or propyl

CH_3CHCH_3
| isopropyl; or 2-propyl

$CH_3CH_2CH_2CH_2$—   n-butyl; or butyl

Note that two different propyl groups are formed, depending on whether the hydrogen atom removed is from an end carbon or from a middle carbon atom. In a like manner, four different butyl groups are formed (only one of which is shown). Alkyl groups are often designated by the letter R−. Thus RH represents an alkane hydrocarbon.

Functional Groups

The formulas for many classes of organic compounds may be derived from the formulas of the alkane hydrocarbons by substituting a different group for one or more of the hydrogen atoms in the hydrocarbon chain. The groups are known as **functional groups** and characterize the classes of compounds that they represent.

The functional group of the alcohols is $-OH$. Two examples are methyl alcohol (CH_3OH) and ethyl alcohol (CH_3CH_2OH or C_2H_5OH). In a like manner, the formulas of an entire series of alcohols may be written by substituting an $-OH$ group for a hydrogen atom on the alkane chain or by combining the $-OH$ group with the alkyl groups given above. Thus n-propyl alcohol and isopropyl alcohol are $CH_3CH_2CH_2OH$ and CH_3CHCH_2, respectively.

$$\qquad\qquad\qquad\qquad | \\ \qquad\qquad\qquad\qquad OH$$

Other common functional groups, together with their classes of compounds, are given below.

Functional Group	Class of Compound	Examples*	
R $-$ H	Alkane hydrocarbon	CH_4	methane
		CH_3CH_3	ethane
$\diagdown C{=}C\diagup$	Alkene hydrocarbon	$CH_2 = CH_2$	Ethene (Ethylene)
		$CH_3CH = CH_2$	Propene (Propylene)
$-OH$	Alcohol	CH_3OH	Methanol (Methyl alcohol)
		CH_3CH_2OH	Ethanol (Ethyl alcohol)
$\overset{O}{\underset{\|\|}{}}$ $-C-OH$ or $-COOH$	Carboxylic acid	HCOOH	Methanoic acid (Formic acid)
		CH_3COOH	Ethanoic acid (Acetic acid)
$\overset{H}{\underset{\|}{}}$ $-C=O$	Aldehyde	$H-\overset{H}{\underset{\|}{C}}=O$	Methanal (Formaldehyde)
		$CH_3\overset{H}{\underset{\|}{C}}=O$	Ethanal (Acetaldehyde)
R$-$C$-$R $\underset{O}{\underset{\|\|}{}}$	Ketone	$CH_3\underset{O}{\underset{\|\|}{C}}CH_3$	Propanone (Acetone)
$\overset{O}{\underset{\|\|}{}}$ $-C-OR$	Ester	$H-\overset{O}{\underset{\|\|}{C}}-OCH_3$	Methyl methanoate (Methyl formate)
		$CH_3-\overset{O}{\underset{\|\|}{C}}-OCH_2CH_3$	Ethyl ethanoate (Ethyl acetate)

*IUPAC name   Common name in parentheses

EXERCISE 1

Significant Figures and Exponential Notation

1. How many significant figures are in each of the following numbers?

   (a) 7.42 _____ (b) 4.6 _____ (c) 3.40 _____ (d) 26,000 _____

   (e) 0.088 _____ (f) 0.0034 _____ (g) 0.0230 _____ (h) 0.3080 _____

2. Write each of the following numbers in proper exponential notation:

   (a) 423   (a) _____

   (b) 0.032   (b) _____

   (c) 8,300   (c) _____

   (d) 302.0   (d) _____

   (e) 12,400,000   (e) _____

   (f) 0.0007   (f) _____

3. How many significant figures should be in the answer to each of the following calculations?

   (a) $\begin{array}{r} 17.10 \\ + 0.77 \\ \hline \end{array}$ (b) $\begin{array}{r} 57.826 \\ -9.4 \\ \hline \end{array}$ (a) _____

   (b) _____

   (c) $12.4 \times 2.82 =$ (d) $6.4 \times 3.1416 =$ (c) _____

   (d) _____

   (e) $\dfrac{0.5172}{0.2742} =$ (f) $\dfrac{0.0172}{4.36} =$ (e) _____

   (f) _____

   (g) $\dfrac{5.82 \times 760. \times 425}{723 \times 273} =$ (h) $\dfrac{0.92 \times 454 \times 5.620}{22.4} =$ (g) _____

   (h) _____

4. For each of these problems, complete the answer with a 10 raised to the proper power. Note that each answer is expressed to the correct number of significant figures.

(a) $2.71 \times 10^4 \times 2.0 \times 10^2 = 5.4 \times$ _____   (a) _____

(b) $\dfrac{4.523 \times 10^4}{2.71 \times 10^2} = 1.67 \times$ _____   (b) _____

(c) $4.8 \times 10^4 \times 3.5 \times 10^4 = 1.7 \times$ _____   (c) _____

(d) $\dfrac{1.64 \times 10^{-4}}{1.2 \times 10^2} = 1.4 \times$ _____   (d) _____

(e) $\dfrac{4.70 \times 10^2}{8.42 \times 10^5} = 5.58 \times$ _____   (e) _____

5. Solve each of the following problems, expressing each answer to the proper number of significant figures. Use exponential notation for (c), (d), and (e).

(a)   1.842   (b) 714.3
        45.21   − 28.52   (a) _____
      + 37.55

                                          (b) _____

(c) $2.83 \times 10^3 \times 7.55 \times 10^7 =$ (c) _____

(d) $4.4 \times 5{,}280 =$ (d) _____

(e) $\dfrac{7.07 \times 10^{-4} \times 6.51 \times 10^{-2}}{2.92 \times 10^4} =$ (e) _____

Answers

1. (a) 3, (b) 2, (c) 3, (d) 2, (e) 2, (f) 2, (g) 3, (h) 4.

2. (a) 4.23×10^2, (b) 3.2×10^{-2}, (c) 8.3×10^3, (d) 3.020×10^2, (e) 1.24×10^7, (f) 7×10^{-4}.

3. (a) 4, (b) 3, (c) 3, (d) 2, (e) 4, (f) 3, (g) 3, (h) 2.

4. (a) 10^6, (b) 10^2, (c) 10^9, (d) 10^{-6}, (e) 10^{-4}.

5. (a) 84.60, (b) 685.8, (c) 2.14×10^{11}, (d) 2.3×10^4, (e) 1.58×10^{-9}.

EXERCISE 2

Measurements

For each of the following problems, show your calculation setup. In both your setup and answer, show units and follow the rules of significant figures. See Experiment 2 and the appendixes for any needed formulas or conversion factors.

1. Convert 78°F to degrees Celsius.

    _____

2. Convert −13°C to degrees Fahrenheit.

    _____

3. An object weighs 8.22 lbs. What is the mass in grams?

    _____

4. A stick is 12.0 cm long. What is the length in inches?

    _____

5. The water in a flask measures 423 mL. How many quarts is this?

    _____

6. A piece of lumber measures 98.4 cm long. What is its length in:

    (a) Millimeters?

    _____

    (b) Feet?

    _____

7. A block of wood has a volume of 35.3 cm³. Its mass is 31.7 g. Calculate the density of the block.

　　　　　　　　　　　　　　　　　　　　　　　　　　　_____

8. A graduated cylinder was filled to 25.0 mL with liquid. A solid object weighing 73.5 g was immersed in the liquid, raising the liquid level to 43.9 mL. Calculate the density of the solid object.

　　　　　　　　　　　　　　　　　　　　　　　　　　　_____

9. The density of the liquid in Problem 8 is 0.874 g/mL. What is the mass of the liquid in the graduated cylinder.

　　　　　　　　　　　　　　　　　　　　　　　　　　　_____

10. How many joules of heat are absorbed by 500.0 g of water when its temperature increases from 20.0°C to 80.0°C? (sp. ht. water = 1.00 cal/g°C)

　　　　　　　　　　　　　　　　　　　　　　　　　　　_____

11. A beaker contains 421 mL of water. The density of the water is 1.00 g/mL. Calculate:

　　(a) The volume of the water in liters.

　　　　　　　　　　　　　　　　　　　　　　　　　　　_____

　　(b) The mass of the water in grams.

　　　　　　　　　　　　　　　　　　　　　　　　　　　_____

12. The density of carbon tetrachloride, CCl_4, is 1.59 g/mL. Calculate the volume of 100.0 g of CCl_4.

　　　　　　　　　　　　　　　　　　　　　　　　　　　_____

NAME ______________________

SECTION __________ DATE __________

EXERCISE 3

INSTRUCTOR ______________________

Names and Formulas 1

Give the names of the following compounds:

1. $NaCl$ ______________________
2. $AgNO_3$ ______________________
3. $BaCrO_4$ ______________________
4. $Ca(OH)_2$ ______________________
5. $ZnCO_3$ ______________________
6. Na_2SO_4 ______________________
7. Al_2O_3 ______________________
8. $CdBr_2$ ______________________
9. KNO_2 ______________________
10. $Fe(NO_3)_3$ (a) ______________________

    (b) ______________________
11. $(NH_4)_3PO_4$ ______________________
12. $KClO_3$ ______________________
13. MgS ______________________
14. $Cu_2C_2O_4$ ______________________

Give the formulas of the following compounds:

1. Barium chloride

2. Zinc fluoride

3. Lead(II) iodide

4. Ammonium hydroxide

5. Potassium chromate

6. Bismuth(III) chloride

7. Magnesium perchlorate

8. Copper(II) sulfate

9. Iron(III) chloride

10. Calcium cyanide

11. Copper(I) sulfide

12. Silver carbonate

13. Cadmium hypochlorite

14. Sodium bicarbonate

15. Aluminum acetate

16. Nickel(II) phosphate

17. Sodium sulfite

18. Tin(IV) oxide

1. _____

2. _____

3. _____

4. _____

5. _____

6. _____

7. _____

8. _____

9. _____

10. _____

11. _____

12. _____

13. _____

14. _____

15. _____

16. _____

17. _____

18. _____

EXERCISE 4

Names and Formulas II

Give the names of the following compounds:

1. $(NH_4)_2S$ _____

2. NiF_2 _____

3. $Sb(ClO_3)_3$ _____

4. $HgCl_2$ _____

5. $H_2SO_4(aq)$ _____

6. $CrBr_3$ _____

7. Cu_2CO_3 (a) _____

    (b) _____

8. $K_2Cr_2O_7$ _____

9. $FeSO_4$ (a) _____

    (b) _____

10. $AgC_2H_3O_2$ _____

11. HCl _____

12. $HCl(aq)$ _____

13. $KBrO_3$ _____

14. $Cd(ClO_2)_2$ _____

Give the formulas of the following compounds:

1. Sodium oxalate   1. _____

2. Manganese(II) iodate   2. _____

3. Zinc nitrite   3. _____

4. Potassium permanganate   4. _____

5. Titanium(IV) bromide   5. _____

6. Sodium arsenate   6. _____

7. Manganese(IV) sulfide   7. _____

8. Bismuth(III) arsenate   8. _____

9. Sodium peroxide   9. _____

10. Magnesium bicarbonate   10. _____

11. Lead(II) acetate   11. _____

12. Phosphoric acid   12. _____

13. Nitric acid   13. _____

14. Acetic acid   14. _____

15. Arsenic(III) iodide   15. _____

16. Ammonium thiocyanate   16. _____

17. Cobalt(II) chlorite   17. _____

18. Stannous fluoride   18. _____

NAME ______________________

SECTION __________ DATE __________

EXERCISE 5

INSTRUCTOR ______________________

Names and Formulas III

Give the names of the following compounds:

1. CO_2 ______________________
2. H_2O_2 ______________________
3. $Ni(MnO_4)_2$ ______________________
4. $Co_3(AsO_4)_2$ ______________________
5. KCN ______________________
6. Sb_2O_5 ______________________
7. BaH_2 ______________________
8. $NaHSO_3$ ______________________
9. $As(NO_2)_5$ ______________________
10. KSCN ______________________
11. Ag_2CO_3 ______________________
12. CrF_3 ______________________
13. SnS_2 (a) ______________________

    (b) ______________________
14. $H_2SO_3(aq)$ ______________________
15. HgC_2O_4 ______________________
16. $Pb(HCO_3)_2$ ______________________
17. $Cu(OH)_2$ ______________________

Give the formulas of the following substances:

1. Ammonium hydrogen carbonate   1. _____

2. Hydrogen sulfide   2. _____

3. Barium hydroxide   3. _____

4. Carbon tetrachloride   4. _____

5. Nickel(II) perchlorate   5. _____

6. Lead(II) nitrate   6. _____

7. Sulfur dioxide   7. _____

8. Carbonic acid   8. _____

9. Copper(I) carbonate   9. _____

10. Calcium cyanide   10. _____

11. Arsenic(III) oxide   11. _____

12. Silver dichromate   12. _____

13. Nitrous acid   13. _____

14. Copper(II) bromide   14. _____

15. Ammonia   15. _____

16. Chlorine   16. _____

17. Chromium(III) sulfite   17. _____

18. Chloric acid   18. _____

19. Barium arsenate   19. _____

20. Manganese(IV) chloride   20. _____

21. Carbon disulfide   21. _____

22. Aluminum fluoride   22. _____

EXERCISE 6

Equation Writing and Balancing I

Balance the following equations:

1. $Mg + O_2 \xrightarrow{\Delta} MgO$

2. $KClO_3 \xrightarrow{\Delta} KCl + O_2$

3. $Fe + O_2 \xrightarrow{\Delta} Fe_3O_4$

4. $Mg + HCl \longrightarrow MgCl_2 + H_2$

5. $Na + H_2O \longrightarrow NaOH + H_2$

Beneath each word equation write the formula equation and balance it. Remember that oxygen and hydrogen are diatomic molecules.

1. Sulfur + Oxygen $\xrightarrow{\Delta}$ Sulfur Dioxide

2. Zinc + Sulfuric acid $\longrightarrow$ Zinc sulfate + Hydrogen

3. Carbon + Oxygen $\xrightarrow{\Delta}$ Carbon dioxide

4. Hydrogen + Oxygen $\xrightarrow{\Delta}$ Water

5. Aluminum + Hydrochloric acid $\longrightarrow$ Aluminum chloride + Hydrogen

Balance the following equations:

1. $N_2 +$ $H_2 \xrightarrow{\Delta}$ NH_3

2. $CoCl_2 \cdot 6H_2O \xrightarrow{\Delta}$ $CoCl_2 +$ H_2O

3. $Fe +$ $H_2O \xrightarrow{\Delta}$ $Fe_3O_4 +$ H_2

4. $F_2 +$ $H_2O \xrightarrow{\Delta}$ $HF +$ O_2

5. $Pb(NO_3)_2 \xrightarrow{\Delta}$ $PbO +$ $NO +$ O_2

Beneath each word equation write and balance the formula equation. Oxygen, hydrogen, and bromine are diatomic molecules

1. Aluminum + Oxygen $\xrightarrow{\Delta}$ Aluminum oxide

2. Potassium + Water $\longrightarrow$ Potassium hydroxide + Hydrogen

3. Arsenic(III) oxide + Hydrochloric acid $\longrightarrow$ Arsenic(III) chloride + Water

4. Phosphorus + Bromine $\longrightarrow$ Phosphorus tribromide

5. Sodium bicarbonate + Nitric acid $\longrightarrow$ Sodium nitrate + Water + Carbon dioxide

EXERCISE 7

Equation Writing and Balancing II

Complete and balance the following double displacement reaction equations (assume all reactions go to products):

1. $NaCl +$ $AgNO_3 \longrightarrow$

2. $BaCl_2 +$ $H_2SO_4 \longrightarrow$

3. $NaOH +$ $HCl \longrightarrow$

4. $Na_2CO_3 +$ $HCl \longrightarrow$

5. $H_2SO_4 +$ $NH_4OH \longrightarrow$

6. $FeCl_3 +$ $NH_4OH \longrightarrow$

7. $Na_2SO_3 +$ $HCl \longrightarrow$

8. $K_2CrO_4 +$ $Pb(NO_3)_2 \longrightarrow$

9. $NaC_2H_3O_2 +$ $HCl \longrightarrow$

10. $NaOH +$ $NH_4NO_3 \longrightarrow$

11. $BiCl_3 +$ $H_2S \longrightarrow$

12. $K_2C_2O_4 +$ $HCl \longrightarrow$

13. $H_3PO_4 +$ $Ca(OH)_2 \longrightarrow$

14. $(NH_4)_2CO_3 +$ $HNO_3 \longrightarrow$

15. $K_2CO_3 +$ $NiBr_2 \longrightarrow$

Complete and balance the following equations. (Combination, 1–4; Decomposition, 5–8; Single displacement, 9–12; Double displacement, 13–16.)

1. $K + Cl_2 \longrightarrow$

2. $Zn + O_2 \longrightarrow$

3. $BaO + H_2O \longrightarrow$

4. $SO_3 + H_2O \longrightarrow$

5. $MgCO_3 \xrightarrow{\Delta}$

6. $NH_4OH \xrightarrow{\Delta}$

7. $Mn(ClO_3)_2 \xrightarrow{\Delta}$

8. $HgO \xrightarrow{\Delta}$

9. $Ni + HCl \longrightarrow$

10. $Pb + AgNO_3 \longrightarrow$

11. $Cl_2 + NaI \longrightarrow$

12. $Al + CuSO_4 \longrightarrow$

13. $KOH + H_3PO_4 \longrightarrow$

14. $Na_2C_2O_4 + CaCl_2 \longrightarrow$

15. $(NH_4)_2SO_4 + KOH \longrightarrow$

16. $ZnCl_2 + (NH_4)_2S \longrightarrow$

EXERCISE 8

Equation Writing and Balancing III

For each of the following situations, write and balance the formula equation for the reaction that occurs.

1. A strip of zinc is dropped into a test tube of hydrochloric acid.

2. Hydrogen peroxide decomposes in the presence of manganese dioxide.

3. Copper(II) sulfate pentahydrate is heated to drive off the water of hydration.

4. A piece of sodium is dropped into a beaker of water.

5. A piece of limestone (calcium carbonate) is heated in a Bunsen burner flame.

6. A piece of zinc is dropped into a solution of silver nitrate.

7. Hydrochloric acid is added to a sodium carbonate solution.

8. Potassium chlorate is heated in the presence of manganese dioxide.

9. Hydrogen gas is burned in air.

10. Sulfuric acid solution is reacted with sodium hydroxide solution.

EXERCISE 9

Graphical Representation of Data

A. From the figure at the right, read values for the following:

1. The vapor pressure of ethyl ether at 20°C.

    _____

2. The temperature at which ethyl chloride has a vapor pressure of 620 torr.

    _____

3. The temperature at which ethyl alcohol has the pressure that ethyl chloride has at 2°C.

    _____

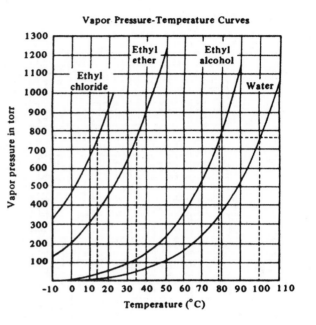

B. Plotting Graphs

1. Plot the following pressure-temperature data for a gas on the graph below. Draw the best possible straight line through the data. Provide temperature and pressure scales.

Temperature, °C	0	20	40	60	80	100
Pressure, torr	586	628	655	720	757	800

Pressure-Temperature Data for a Gas

Pressure, torr

Temperature, °C

2. (a) Study the data given below; (b) determine suitable scales for pressure and for volume and mark these scales on the graph; (c) plot eight points on the graph; (d) draw the best possible line through these points; (e) place a suitable title at the top of the graph.

Pressure-volume data for a gas

Volume, mL	10.70	7.64	5.57	4.56	3.52	2.97	2.43	2.01
Pressure, torr	250	350	480	600	760	900	1100	1330

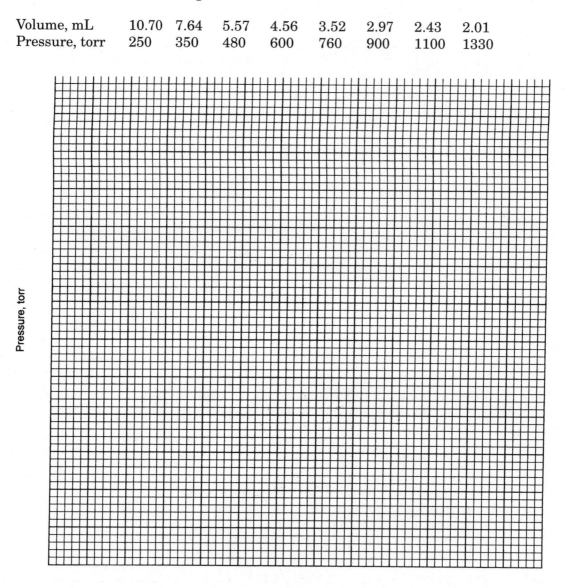

Pressure, torr

Volume, mL

Read from your graph:

(a) The pressure at 10.0 mL _____

(b) The volume at 700 torr _____

NAME ______________________

SECTION __________ DATE __________

EXERCISE 10

INSTRUCTOR ______________________

Moles

Show calculation setups and answers for all problems.

1. Find the molar mass of (a) nitric acid, HNO_3; (b) potassium bicarbonate, $KHCO_3$; and (c) Nickel(II) nitrate, $Ni(NO_3)_2$.

(a) ______________

(b) ______________

(c) ______________

2. A sample of mercury(II) bromide, $HgBr_2$, weighs 8.65 g. How many moles are in this sample?

____

3. What is the mass of 0.45 mol of ammonium sulfate, $(NH_4)_2SO_4$?

____

4. How many molecules are contained in 6.53 mol of nitrogen gas, N_2?

____

5. Calculate the percent composition by mass of calcium sulfite, $CaSO_3$.

Ca ______________

S ______________

O ______________

6. An organic compound is analyzed and found to be carbon 51.90%, hydrogen 9.80%, and chlorine 38.30%. What is the empirical formula of this compound?

_____

7. A sample of oxygen gas, O_2, weighs 28.4 g. How many molecules of O_2 and how many atoms of O are present in this sample?

_____ molecules of O_2

_____ atoms of O

8. A mixture of sand and salt is found to be 48 percent NaCl by mass. How many moles of NaCl are in 74 g of this mixture?

_____

9. What is the mass of 2.6×10^{23} molecules of ammonia, NH_3?

_____

10. A water solution of sulfuric acid has a density of 1.67 g/mL and is 75 percent H_2SO_4 by mass. How many moles of H_2SO_4 are contained in 400. mL of this solution?

_____

NAME ____________________

SECTION __________ DATE __________

EXERCISE 11

INSTRUCTOR ____________________

Stoichiometry I

Show calculation setups and answers for all problems.

1. Use the equation given to solve the following problems:

$$Na_3PO_4 + 3\,AgNO_3 \longrightarrow Ag_3PO_4 + 3\,NaNO_3$$

(a) How many moles of Na_3PO_4 would be required to react with 1.0 mol of $AgNO_3$?

____

(b) How many moles of $NaNO_3$ can be produced from 0.50 mol of Na_3PO_4?

____

(c) How many grams of Ag_3PO_4 can be produced from 5.00 g of Na_3PO_4?

____

(d) If you have 9.44 g of Na_3PO_4, how many grams of $AgNO_3$ will be needed for complete reaction?

____

(e) When 25.0 g of $AgNO_3$ are reacted with excess Na_3PO_4, 18.7 g of Ag_3PO_4 are produced. What is the percentage yield of Ag_3PO_4?

____

2. Use the equation given to solve the following problems:

$$2\,KMnO_4 + 16\,HCl \longrightarrow 5\,Cl_2 + 2\,KCl + 2\,MnCl_2 + 8\,H_2O$$

(a) How many moles of HCl are required to react with 35 g of $KMnO_4$?

_____

(b) How many Cl_2 molecules will be produced using 3.0 mol $KMnO_4$?

_____

(c) To produce 35.0 g of $MnCl_2$, what mass of HCl will need to react?

_____

(d) How many moles of water will be produced when 8.0 mol of $KMnO_4$ react?

_____

(e) What is the maximum mass of Cl_2 that can be produced by reacting 70.0 g of $KMnO_4$ with 15.0 g of HCl?

_____

NAME ____________________

SECTION __________ DATE __________

EXERCISE 12

INSTRUCTOR ____________________

Gas Laws

Show calculation setups and answers for all problems.

1. A sample of nitrogen gas, N_2, occupies 3.0 L at a pressure of 3.0 atm. What volume will it occupy when the pressure is changed to 0.50 atm and the temperature remains constant?

____

2. A sample of methane gas, CH_4, occupies 4.50 L at a temperature of 20.0°C. If the pressure is held constant, what will be the volume of the gas at 100.°C?

____

3. The pressure of hydrogen gas in a constant-volume cylinder is 4.25 atm at 0°C. What will the pressure be if the temperature is raised to 80°C?

____

4. A 325 mL sample of air is at 720. torr and 30.0°C What volume will this gas occupy at 800. torr and 75.0°C?

____

5. A sample of gas occupies 500. mL at STP What volume will the gas occupy at 85.0°C and 525 torr?

____

6. A quantity of oxygen occupies a volume of 19.2 L at STP. How many moles of oxygen are present?

_____

7. A 425 mL volume of hydrogen chloride gas, HCl, is collected at 25°C and 720. torr. What volume will it occupy at STP?

_____

8. What volume would 10.5 g of nitrogen gas, N_2, occupy at 200. K and 2.02 atm?

_____

9. Calculate the density of sulfur dioxide, SO_2, at STP.

_____

10. In a laboratory experiment, 133 mL of gas was collected over water at 24°C and 742 torr. Calculate the volume that the dry gas would occupy at STP.

_____

11. A volume of 122 mL of argon, Ar, is collected at 50°C and 758 torr. What does this sample weigh?

_____

NAME ____________________

SECTION __________ DATE __________

EXERCISE 13

INSTRUCTOR ____________________

Solution Concentrations

Show calculation setups and answers for all problems.

1. What will be the percent composition by mass of a solution made by dissolving 15.0 g of barium nitrate, $Ba(NO_3)_2$, in 45.0 g of water?

$Ba(NO_3)_2$ ____________________

H_2O ____________________

2. How many moles of potassium hydroxide, KOH, are required to prepare 2.00 L of 0.250 M solution?

____

3. What is the molarity of a solution that contains 3.50 g of sodium hydroxide, NaOH in 150. mL of solution?

____

4. How many milliliters of 0.400 M solution can be prepared from 5.00 g of NaBr?

____

5. How many grams of potassium bromide, KBr, could be recovered by evaporating 650. mL of 15.0 percent KBr solution to dryness (d = 1.11g/mL)?

____

6. How many milliliters of 12.0 M HCl is needed to prepare 300. mL of 0.250 M HCl solution?

_____

7. A sample of potassium hydrogen oxalate, KHC_2O_4, weighing 0.717 g, was dissolved in water and titrated with 23.47 mL of an NaOH solution. Calculate the molarity of the NaOH solution.

_____

8. How many grams of hydrogen chloride are in 50. mL of concentrated (12 M) HCl solution?

_____

9. A sulfuric acid solution has a density of 1.49 g/mL and contains 59 percent H_2SO_4 by mass. What is the molarity of this solution?

_____

10. Sulfuric acid reacts with sodium hydroxide according to this equation:

$$H_2SO_4 + 2\,NaOH \longrightarrow Na_2SO_4 + 2\,H_2O$$

A 10.00 mL sample of the H_2SO_4 solution required 18.71 mL of 0.248 M NaOH for neutralization. Calculate the molarity of the acid.

_____

NAME ____________________

SECTION __________ DATE __________

EXERCISE 14

INSTRUCTOR ____________________

Stoichiometry II

Show calculation setups and answers for all problems.

1. Use the equation to solve the following problems:

$$6\,KI + 8\,HNO_3 \longrightarrow 6\,KNO_3 + 2\,NO + 3\,I_2 + 4\,H_2O$$

(a) When 38 g of KI are reacted, how many grams of I_2 will be formed?

____

(b) What volume of NO gas, measured at STP, will be produced when 47.0 g of HNO_3 are reacted?

____

(c) How many milliliters of 6.00 M HNO_3 will react with 1.00 mole of KI?

____

(d) When the reaction produces 8.0 mol of NO, how many molecules of I_2 will be produced?

____

(e) How many grams of iodine can be obtained by reacting 35.0 mL of 0.250 M KI solution?

____

2. Use the equation given to solve the following problems. All substances are in the gas phase.

$$N_2(g) + 3\,H_2(g) \longrightarrow 2\,NH_3(g)$$

   (a) When 2.0 mol of H_2 react, how many moles of NH_3 will be formed?

   _____

   (b) When 5.50 mol of N_2 react, what volume of NH_3, measured at STP, will be formed?

   _____

   (c) What volume of NH_3 will be formed when 12.0 L of H_2, are reacted? All volumes are measured at STP

   _____

   (d) How many molecules of NH_3 will be formed when 30.0 L of N_2 at STP react?

   _____

   (e) What volume of NH_3, measured at 25°C and 710. torr, will be produced from 18.0 g of H_2?

   _____

   (f) When a mixture of 9.00 L of N_2 and 30.0 L of H_2 are reacted, what volume of NH_3 can be produced? Assume STP conditions.

   _____

NAME ____________________

SECTION __________ DATE __________

EXERCISE 15

INSTRUCTOR ____________________

Chemical Equilibrium

1. Consider the following system at equilibrium:

$$2\,CO_2(g) + 135.2\,\text{kcal} \rightleftharpoons 2\,CO(g) + O_2(g)$$

Complete the following table. Indicate changes in moles and concentrations by entering I, D, N, or? in the table (I = increase, D = decrease, N = no change, ? = insufficient information to determine).

Change or stress imposed on the system at equilibrium	Direction of shift, left or right, to reestablish equilibrium	Change in number of moles			Change in molar concentrations		
		CO_2	CO	O_2	CO_2	CO	O_2
a. Add CO							
b. Remove CO_2							
c. Decrease volume of reaction vessel							
d. Increase temperature							
e. Add catalyst							
f. Add both CO_2 and O_2							

2. Consider the reaction $PCl_5(g) \rightleftharpoons PCl_3(g) + Cl_2(g)$

At 250°C, PCl_5 is 45% decomposed.

(a) When 0.110 mol of PCl_5(g) is introduced into a 1.00 L container at 250°C, what will be the equilibrium concentrations of PCl_5, PCl_3, and Cl_2?

PCl_5 ____________________

PCl_3 ____________________

Cl_2 ____________________

(b) What is the value of K_{eq} at 250°C?

K_{eq} ____________________

3. For the reaction $H_2(g) + I_2(g) \rightleftharpoons 2\,HI(g)$, $K_{eq} = 0.17$ at 500 K. What concentration of $I_2(g)$ will be in equilibrium with $H_2 = 0.040$ M, $HI = 0.015$ M?

_____

4. $CaCO_3$ has a solubility in water of 6.9×10^{-5} mol/L. Calculate the solubility product constant.

_____

5. A 0.40 M HClO solution was found to have an H^+ concentration of 1.1×10^{-4} M. Calculate the value of the ionization constant. The ionization equation is $HClO \rightleftharpoons H^+ + ClO^-$.

_____

6. Calculate (a) the H^+ ion concentration, (b) the pH, and (c) the percent ionization of a 0.40 M solution of $HC_2H_3O_2$ ($K_a = 1.8 \times 10^{-5}$).

(a) _____

(b) _____

(c) _____

EXERCISE 16

Oxidation-Reduction Equations I

Balance the following oxidation-reduction equations:

1. $P +$ $HNO_3 +$ $H_2O \longrightarrow$ $H_3PO_4 +$ NO

2. $H_2SO_4 +$ $HI \longrightarrow$ $H_2S +$ $I_2 +$ H_2O

3. $KBrO_2 +$ $KI +$ $HBr \longrightarrow$ $KBr +$ $I_2 +$ H_2O

4. $Sb +$ $HNO_3 \longrightarrow$ $Sb_2O_5 +$ $NO +$ H_2O

5. $NO_2 +$ $H_2O \longrightarrow$ $HNO_3 +$ NO

6. Br_2 + NH_3 $\longrightarrow$ NH_4Br + N_2

7. KI + HNO_3 $\longrightarrow$ KNO_3 + NO + I_2 + H_2O

8. H_2SO_3 + $KMnO_4$ $\longrightarrow$ $MnSO_4$ + H_2SO_4 + K_2SO_4 + H_2O

9. $K_2Cr_2O_7$ + H_2O + S $\longrightarrow$ SO_2 + KOH + Cr_2O_3

10. $KMnO_4$ + HCl $\longrightarrow$ Cl_2 + KCl + $MnCl_2$ + H_2O

EXERCISE 17

Oxidation-Reduction Equations II

Balance the following oxidation-reduction equations using the ion-electron method.

1. $MnO_4^- + Cl^- + H^+ \longrightarrow Mn^{2+} + Cl_2 + H_2O$

2. $Ag_2S + NO_3^- + H^+ \longrightarrow S + NO + Ag^+ + H_2O$

3. $ClO_4^- + I^- + H^+ \longrightarrow I_2 + Cl^- + H_2O$

4. $Br_2 + H_2O \longrightarrow BrO_3^- + Br^- + H^+$

5. $MnO_4^- + HS^- + H_2O \longrightarrow S + MnO_2 + OH^-$

6. $H_2O_2 + IO_3^-$ $\longrightarrow$ $I^- +$ O_2 (acid solution)

7. $Cl_2 +$ $SO_2 \longrightarrow$ $SO_4^{2-} +$ Cl^- (acid solution)

8. $U^{4+} +$ $MnO_4^- \longrightarrow$ $Mn^{2+} +$ UO_2^{2+} (acid solution)

9. $Fe(CN)_6^{3-} +$ $Cr_2O_3 \longrightarrow$ $Fe(CN)_6^{4-} +$ CrO_4^{2-} (basic solution)

10. $Cr(OH)_3 +$ $O_2^{2-} \longrightarrow$ CrO_4^{2-} (basic solution)

EXERCISE 18

Organic Chemistry I

1. Write structural formulas for the three different isomers of pentane, all having the molecular formula C_5H_{12}.

2. Write condensed structural formulas for (a) ethene, (b) propene, (c) three isomers of butene, (d) isomers of pentene, C_5H_{10}.

   (a)   (b)   (c)

   (d)

3. Write a balanced equation for the complete combustion of ethane.

4. Write condensed structural formulas for the five isomers of hexane, all having the molecular formula C_6H_{14}.

5. Write condensed structural formulas for (a) acetylene, (b) 2-methylpropane (c) benzene, (d) *n*-octane

   (a)   (b)

   (c)   (d)

6. Write a balanced equation for the complete combustion of benzene.

EXERCISE 19

Organic Chemistry II

1. Write condensed structural formulas for (a) ethyl alcohol, (b) acetone, (c) formaldehyde, (d) acetic acid.

   (a)   (b)

   (c)   (d)

2. Write condensed structural formulas for the four different isomers of butyl alcohol, all having the formula C_4H_9OH.

3. Write the name of the ester that can be derived from the following pairs of acids and alcohols:

Alcohol	Acid	Ester
Isopropyl alcohol	Acetic acid	
Ethyl alcohol	Salicylic acid	
Methyl alcohol	Stearic acid	

4. The formula of butanoic acid is $CH_3CH_2CH_2COOH$. Write the structural formula for methyl butanoate.

5. The formula for benzoic acid is ⬡—COOH

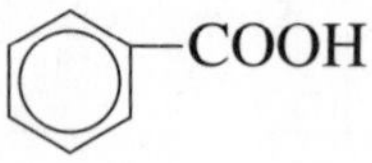

   Write the structural formula for methyl benzoate.

6. Identify the class of compound for each of the following:

(a)　　　O
　　　　‖
　　　CH₃C—OH

(b) CH₃CHCH₃
　　　　|
　　　　OH

(a) _____

(b) _____

(c)　　　O
　　　　‖
　　　CH₃C—OCH₃

(d) ⬡—COOH

(c) _____

(d) _____

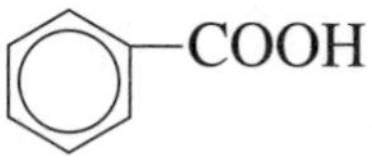

(e)　　　　　　O
　　　　　　　‖
　CH₃—CH₂—C—CH₃

(f)　　　　H
　　　　　|
　　CH₃CHC=O
　　　　|
　　　　CH₃

(e) _____

(f) _____

(g) 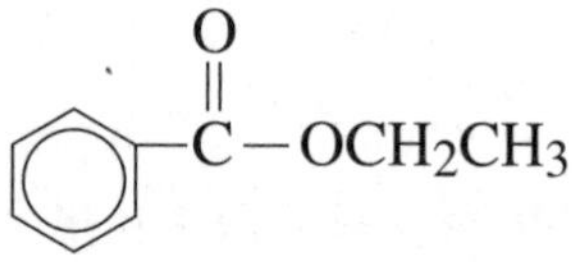　　　　O
　　　　　‖
　⬡—C—OCH₂CH₃

(h)　　　CH₃
　　　　　|
　　CH₃CHCH₂CH₃

(g) _____

(h) _____

(i) CH₃COOH

(j)　　　O
　　　　‖
　CH₃C—OCH₃

(i) _____

(j) _____

(k) CH₃CH(OH)CH₃

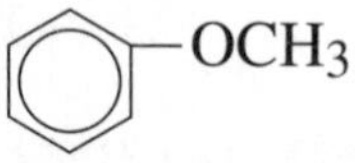

(l) ⬡—OCH₃

(k) _____

(l) _____

APPENDIX 1

Suggested List of Equipment

Equipment for Student Lockers

1. 5 Beakers: 50, 100, 150, 250, 400 mL
2. 1 Burner, Tirrill (optional)
3. Ceramfab pad
4. 1 Clay triangle
5. 2 Crucibles, size 0
6. 2 Crucible covers, size F
7. 1 Crucible tongs
8. 1 Evaporating dish, size 1
9. 1 File, triangular
10. 1 Filter paper (box)
11. 2 Flasks, Erlenmeyer, 125 mL
12. 2 Flasks, Erlenmeyer, 250 mL
13. 1 Flask, Florence, 500 mL
14. 5 Glass plates, 3 × 3 in.
15. 1 Graduated cylinder, 10 mL
16. 1 Graduated cylinder, 50 mL
17. 2 Litmus paper (vials), red and blue
18. 2 Medicine droppers/disposable pipets
19. 1 Pipet, volumetric, 10 mL
20. 8 Rubber stoppers: 3 No. 1, solid; 1 No. 1, 1-hole; 1 No. 4, 1-hole; 1 No. 4, 2-hole; 1 No. 5, solid; 1 No. 6, 2-hole
21. 2 Rubber tubing (about 25 cm), 3/16 in. diameter
22. 1 Screw clamp
23. 1 Sponge
24. 1 Spatula
25. 12 Test tubes, 18 × 150 mm (or culture tubes)
26. 1 Test tube, ignition, 25 × 200 mm
27. 1 Test tube brush
28. 1 Test tube holder, wire
29. 1 Test tube rack
30. 1 Thermometer, 110°C
31. 1 Thistle top, plastic
32. 1 Utility clamp (single buret clamp)
33. 1 Wash bottle (plastic)
34. 2 Watch glasses, 4 in.
35. 5 Wide-mouth bottles, 8 oz
36. 1 Wing top
37. 1 Wire gauze

Auxillary Equipment Not Supplied in Student Lockers

1. Balances
2. Burners, Tirrill (if not individually supplied)
3. Buret, 25 or 50 mL
4. Deflagration spoon
5. Glass rod, 5 or 6 mm
6. Glass tubing, 6 mm and 8 mm
7. Metric rulers
8. Pipet, volumetric, 10 mL (if not individually supplied)
9. Pneumatic trough
10. Ring stand
11. Ring support, 4 to 5 in. diameter
12. Suction bulbs for pipets
13. pH 1–14 indicator strips
14. Büchner funnel-vacuum flask setup
15. Styrofoam cups, 6 oz.
16. pH meters

APPENDIX 2

List of Reagents Required and Preparation of Solutions

Solids

Ammonium chloride, NH_4Cl

Barium chloride, $BaCl_2 \cdot 2\,H_2O$

Barium sulfate, $BaSO_4$

Benzoic acid, C_6H_5COOH

Boiling chips

Candles

Calcium carbide, CaC_2 (lumps)

Calcium hydroxide, $Ca(OH)_2$

Calcium oxide, CaO

Cardboard squares, 4 inch, with hole in center for thermometer

Cobalt chloride paper

Copper strips, Cu

Copper wire, (No. 18 and 24) Cu

Copper(II) sulfate pentahydrate, $CuSO_4 \cdot 5\,H_2O$

Cotton

Food coloring; red and green paste

Ice

Ice cubes, dark blue

Iron wire (20–24 gauge), Fe

Lead strips, Pb

Lead(II) iodide, PbI_2

Magnesium strips, Mg

Magnesium oxide, MgO

Manganese dioxide, MnO_2

Marble chips, $CaCO_3$

Methylene blue, powder blue

pH indicator strips, 1–14

Potassium bicarbonate, C.P. $KHCO_3$

Potassium chlorate, C.P., $KClO_3$

Potassium chloride, C.P., KCl

Potassium hydrogen phthalate, C.P., $KHC_8H_4O_4$

Salicylic acid, $C_6H_4(COOH)(OH)$

Sand paper or emery cloth

Sodium, Na

Sodium bicarbonate, $NaHCO_3$

Sodium chloride (coarse crystals), $NaCl$

Sodium chloride (fine crystals), $NaCl$

Sodium nitrate, $NaNO_3$

Sodium peroxide, Na_2O_2

Sodium sulfate, Na_2SO_4

Sodium sulfite, Na_2SO_3

Steel wool, Fe (Grade 0 or 1)

Sucrose, $C_{12}H_{22}O_{11}$

Sulfur, S

Tin (II) chloride, $SnCl_2 \cdot 6\,H_2O$

Wood splints

Zinc, mossy, Zn

Zinc strips, Zn (0.01 inch thick)

Zinc sulfate, $ZnSO_4 \cdot 7\,H_2O$

Pure Liquids

Acetone, CH_3COCH_3

Bromine, Br_2, (specimen sample in sealed vial)

Decane, $C_{10}H_{22}$

Ethyl alcohol (ethanol), 95% C_2H_5OH

Ethyl alcohol, denatured anhydrous

Glycerol, $C_3H_5(OH)_3$

Appendix 2 (continued)

Heptane (or low boiling petroleum ether), C_7H_{16} Mineral oil

Isoamyl alcohol (3-methyl-1-butanol), $C_5H_{11}OH$ Pentene (amylene), C_5H_{10}

Isopropyl alcohol (2-propanol), C_3H_7OH Toluene $C_6H_5CH_3$

Kerosene (alkene free)   1,1,1-Trichloroethane, CCl_3CH_3

Methyl alcohol (methanol), CH_3OH

Solutions

All solutions, except where otherwise directed, are prepared by dissolving the designated quantity of solute in distilled water and diluting to 1 liter.

Acetic acid, concentrated (glacial), concentrated reagent

Acetic acid, dilute, 6 M; 350 mL concentrated $HC_2H_3O_2$/liter

Ammonium chloride, 0.1 M; 5.4 g NH_4Cl/liter

Ammonium chloride, saturated; 410 g NH_4Cl/liter

Ammonium hydroxide, concentrated; 15 M reagent, NH_4OH

Ammonium hydroxide, dilute, 6 M; 400 mL concentrated 15 M NH_4OH/liter

Barium chloride, 0.10 M; 24.4 g $BaCl_2 \cdot 2\ H_2O$/liter

Barium hydroxide, saturated; 50 g $Ba(OH)_2 \cdot 8\ H_2O$/liter

Bromine in 1,1,1-trichloroethane, 5% solution; 2.5 mL Br_2 plus 100 mL CCl_3CH_3

Calcium chloride, 0.1 M; 14.7 g $CaCl_2 \cdot 2\ H_2O$/liter

Chlorine water; dilute 150 mL of 5.25% $NaOCl$ (household bleach) to 1 liter. Add 15 mL concentrated HCl and mix gently.

Cobalt (II) chloride, 0.1 M; 23.8 g $CoCl_2 \cdot 6\ H_2O$/liter

Copper (II) nitrate, 0.1 M; 24.2 g $Cu(NO_3)_2 \cdot 3\ H_2O$/liter

Copper (II) sulfate, 0.1 M; 25.0 g $CuSO_4 \cdot 5\ H_2O$/liter

Ethanol/water (50/50 mixture), 500 mL ethanol plus 500. mL water

Formaldehyde, 10% solution; 25 mL formalin (40%) plus 75 mL H_2O

Glucose 0.020 M; 3.6 g $C_6H_{12}O_6$ g/liter

Glucose, 10% solution; 10 g $C_6H_{12}O_6$ plus 90 mL H_2O

Hydrochloric acid, concentrated, concentrated reagent, HCl

Hydrochloric acid, 0.020 M; 6.7 mL of 3.0 M/1 liter

Hydrochloric acid, 1.0 M; 86 mL conc HCl/liter

Hydrochloric acid, dilute, 3 M; 250 mL concentrated HCl/liter

Hydrochloric acid, dilute, 6 M; 500 mL concentrated HCl/liter

Hydrogen peroxide, 3%; 3% reagent solution or 100 mL 30%/liter

Hydrogen peroxide, 9%; 300 mL 30% H_2O_2/liter

Hydrogen peroxide, 30% H_2O_2 (for dilution) (handle with gloves)

Appendix 2 (continued)

Iodine water, saturated; 5 g I_2/liter

Iodine 0.020 M in KI; 1.5 g KI/100 mL, add 0.51 g I_2

Iron (III) chloride, 0.1 M; 27.1 g $FeCl_3 \cdot 6\ H_2O$ + 5 mL concentrated HCl/liter

Lead (II) nitrate, 0.1 M; 33.1 g $Pb(NO_3)_2$/liter

Magnesium sulfate, 0.1 M; 24.6 g $MgSO_4 \cdot 7\ H_2O$/liter

Nickel nitrate, 0.1 M; 29.1 g $Ni(NO_3)_2 \cdot 6\ H_2O$/liter

Nitric acid, concentrated; concentrated reagent, HNO_3

Nitric acid, dilute, 3 M; 188 mL concentrated HNO_3/liter

Nitric acid, dilute, 6 M; 375 mL concentrated HNO_3/liter

Phenolphthalein, 0.2% solution; dissolve 1 g phenolphthalein in 300 mL ethanol (95%) and dilute with water to 500 mL

Phosphoric acid, dilute, 3 M; 201 mL 85% H_3PO_4 solution/liter

Potassium chloride, saturated; 390 g KCl/liter

Potassium nitrate, 0.1 M; 10.1 g KNO_3/liter

Potassium permanganate, 0.1 M; 15.8 g $KMnO_4$/liter

Potassium permanganate, 0.002 M; 0.32 g $KMnO_4$/liter

Potassium thiocyanate, 0.1 M; 9.7 g KSCN/liter

Silver nitrate, 0.10 M; 17.0 g $AgNO_3$/liter

Sodium bromide, 0.1 M; 10.3 g NaBr/liter

Sodium carbonate, 0.020 M; 2.1 g Na_2CO_3/liter

Sodium carbonate, 0.1 M; 10.6 g Na_2CO_3/liter

Sodium chloride, 0.020 M; 1.2 g NaCl/liter

Sodium chloride, 0.1 M; 5.85 g NaCl/liter

Sodium chloride, saturated; 400 g NaCl/liter

Sodium hydroxide, 0.020 M; 0.80 g NaOH/liter

Sodium hydroxide, 1.25 M; 50. g NaOH/liter

Sodium hydroxide, 10% solution; 111 g NaOH/liter

Sodium iodide, 0.1 M; 15.0 g NaI/liter

Sodium phosphate, 0.1 M; 38.0 g $Na_3PO_4 \cdot 12\ H_2O$/liter

Sodium sulfate, 0.1 M; 14.2 g Na_2SO_4/liter

Sulfuric acid, concentrated; concentrated reagent 18 M H_2SO_4

Sulfuric acid, 9 M; 500 mL concentrated H_2SO_4/liter

Sulfuric acid, dilute, 3 M; 167 mL concentrated H_2SO_4/liter

Vinegar, commercial white vinegar, $HC_2H_3O_2$

Wine, commercial red

Zinc nitrate, 0.1 M; 29.8 g $Zn(NO_3)_2$/liter

APPENDIX 3

Special Equipment or Preparations Needed

Experiment 1.   Laboratory Techniques

A small sample of solid lead(II) iodide and sodium nitrate are needed for comparison purposes only.

Experiment 2.   Measurements

An assortment of metal slugs or other solid objects are needed as unknowns for density determination. The diameter of the slugs should be such that they will fit into the 50 mL graduated cylinder. Suggested materials are aluminum, brass, magnesium, steel, etc. Slugs should be numbered for identification.

Experiment 3.   Preparation and Properties of Oxygen

Three demonstrations are suggested (see experiment for details). Büchner funnel-vacuum flask setup for disposal of waste MnO_2.

Experiment 4.   Preparation and Properties of Hydrogen

For safety: Instructor should dispense sodium metal (size of pieces should be no larger than a 4 mm cube).

Experiment 5.   Calorimetry and Specific Heat

An assortment of metal objects like those used for the density determinations in Exp. 2 are needed. They must be small enough to fit into the test tube with id = 22 mm. Styrofoam cups and cardboard cut into 4" squares with a small thermometer hole in the middle should also be available.

Experiment 6.   Freezing Points—Graphing of Data

Slotted corks or stoppers, crushed ice

Experiment 7.   Water in Hydrates

An assortment of samples for unknowns for determination of percent water is needed. Samples can be issued in small coin envelopes or plastic vials. See the Instructor's Manual for the suggested list of samples.

Experiment 8.   Water, Solutions, and pH

The dark blue ice cubes are made by adding methylene blue to tap water until the color is a deep blue. The resulting solution is then frozen in an ice cube tray (Station A6). The green and red colored water is made by adding the food coloring paste to tap water until the resulting solution is brightly colored (Station A4). Much more red water will be needed than green water. The five lengths of capillary tubing needed should have inner diameter measurements that are different depending on what is available. Example: five tubes with i.d. from among the following: 1.0 mm, 1.5 mm, 2.0 mm, 2.5 mm, 3.0 mm, 3.5 mm; or 0.5 mm, 1.0mm, 1.25 mm, 1.75 mm, 2.75 mm; small electric table fan, 1000 mL beaker or battery jar; micropipettes (Pipetman), 200 µl and 1000 µl with disposable tips.

A series of stations is set up for Sections A1-7 and B 3 #6 instead of each student setting up each activity separately at their own lab bench. A station for measuring pH with a pH meter (B3 #6) is recommended. The other B sections of the experiment can be completed by all students at their own place on the bench. The stations needed are:

Station #	Title of the Experiment/ Observation	Materials needed
Station Al	Molecular Structure & Polarity of Water	Ball and stick molecular model kits (2 or 3 should suffice)
Station A2	Polarity of water and Hydrogen Bonds Between Water Molecules	Buret filled with water, 250 mL beaker, plastic rod (even a smooth plastic ruler will work) and soft rayon or silk cloth.
Station A3	Cohesion and Surface Tension	Clean glass microscope slides, dropping bottles of distilled water and 95% ethanol, and liquid detergent; culture dishes, forceps, common pins
Station A4	Capillarity, Cohesion and Adhesion	5 in. pieces of glass capillary tubing with increasing i.d. measurements taped to a white index card with at least 1 inch of the tubes extending beyond the bottom edge of the card; a shallow dish of green water, metric rulers.
Station A5 #1-5	Specific Heat (may want to provide three or four of this station since it takes more time than the other activities)	Two 250. mL Erlenmeyer flasks w/2-hole rubber stoppers inserted; thermometers inserted into one hole; large hot plate; two 600 mL beakers
Station A5, #6	Heat of Vaporization	Two thermometers, riag stand, buret clamp, cotton or rayon tubing to cover the thermometer bulbs; table fan
Station A6.	Water Temperature and Density	1000 mL beaker or battery jar, warm tap water, cold red tap water (pre-mixed, refrigerated to 4°C) 10 mL graduated pipet, pipet pump, blue ice cubes, tongs or gloves

(continued)

Station A7.	Density and volume	Electronic balance with at least 0.001 g precision, weigh boats (1"), micropipettes (Pipetman, 200 μL and 1000 μl), disposable pipet tips
Station B 3 #6	Measurement of pH using a pH Meter	Several pH meters so several students can measure pH for these solutions at the same time. If students are unfamiliar with the use of a pH meter, a card with instructions for use should be with each instrument

Experiment 13.   Ionization—Electrolytes and pH

Conductivity apparatus is needed for the demonstration. The procedure is based on the apparatus described in the experiment but other types may be used without detracting from the results of the demonstration. A magnetic stirrer greatly facilitates the last part of the demonstrations. Two or three pH meters are recommended for student use, set up at stations with the solutions described in the experiment.

Experiment 14.   Identification of Selective Anions

Two unknown solutions (in test tubes) are to be issued to each student. Stock reagents used in the experiment are satisfactory for unknowns.

Experiment 16.   Electromagnetic Energy and Spectroscopy

Hand-held spectroscopes, 1.75 m springs for simulating wave motion (1 per 5 students), vapor lamps with power supplies (2 hydrogen and 2 neon); spectrum chart, incandescent and fluorescent lights, spectrophotometers with range from 350-700 nm, colored pencils, meter sticks, stopwatches (recommended). See Instructor's manual for more details.

Experiment 17.   Lewis Structures and Molecular Models

Ball-and-stick molecular model sets. Two students can share one kit. The number of sets required depends on how many labs are run simultaneously. It is also possible to purchase a large class set of components and divide them into smaller custom kits.

Experiment 18.   Boyle's Law

Boyle's law apparatus is needed. The kits for this experiment can be purchased from several vendors as "Simple Form Boyle's Law Apparatus" or "Elasticity of Gases Kit." The kits include the silicone grease but not the applied weights and vernier calipers. Slotted masses of 0.5 kg and 1 kg allow the applied weights to lie flat on the platform. If not enough slotted masses are available, a combination of bricks and slotted masses works well. One balance per laboratory with the capacity for weighing the heaviest mass to three significant figures. All masses can be preweighed and labeled with tape displaying their mass.

Experiment 20.   Liquids—Vapor Pressure and Boiling Points

125 ml flasks containing acetone, methanol, ethanol, and water are needed for Part A. It is suggested that students work in pairs in Part B. A 1-gallon metal can is needed for the demonstration in Part C.

Experiment 21.   Molar Volume of a Gas

A 3.0 cc or 5.0 cc disposable syringe is needed for each setup. Needle–rubber stopper assemblies that contain a rubber stopper and syringe needle should be preassembled and checked out and in by students. An additional safety feature is to snip off the end of the needle with a wire cutter after it is in the stopper. The needles need to be heavy enough to push through a rubber stopper without bending. 2 L beakers or battery jars. Büchner funnel-vacuum flask setup for disposal of waste MnO_2.

Experiment 22.   Neutralization—Titration I

The following are needed by each student: A small vial or test tube containing about 4 grams of potassium hydrogen phthalate (KHP) (these vials are collected for reuse), one 25 or 50 mL buret, a buret clamp, and 250 mL of NaOH solution of unknown molarity. (See Instructor's Manual for details). The NaOH solution is used in Experiments 22 and 23.

Experiment 23.   Neutralization—Titration II

The following are needed by each student: A 10 mL volumetric pipet, one 25 or 50 mL buret, 50 mL of an acid solution of unknown molarity (See Instructor's Manual for details), 50 mL of vinegar, and 125 mL of standard NaOH solution if Experiment 22 is not done.

Experiment 25.   Heat of Reaction

Styrofoam cups are needed.

Experiment 26.   Distillation of Volatile Liquids

Distillation setup using a 125 mL or 250 mL flask (see Figure 26.1 in the experiment); hot plates or heating mantles (with rheostats). Red wine as the alcoholic beverage for distillation. Pot holders or mitts to handle hot plate.

Experiment 28.   Alcohols, Esters, Aldehydes, and Ketones

Furnish No. 18 copper wire with five or six spiral turns at one end. Wire should be about 20 cm overall in length.

APPENDIX 4

Units of Measurements

Numerical Value of Prefixes with Units

Prefix	Symbol	Number	Power of 10
mega	M	1,000,000	1×10^6
kilo	k	1,000	1×10^3
hecto	h	100	1×10^2
deca	da	10	1×10^1
deci	d	0.1	1×10^{-1}
centi	c	0.01	1×10^{-2}
milli	m	0.001	1×10^{-3}
micro	μ	0.000001	1×10^{-6}
nano	n	0.000000001	1×10^{-9}

Conversion of Units

1 m	=	1000 mm
1 cm	=	10 mm
2.54 cm	=	1 in.
453.6 g	=	1 lb
1 kg	=	2.2 lb, 1000 g
1 g	=	1000 mg
1 L	=	1000 mL
1 mL	=	1 cm^3
0.946 L	=	1 qt
1 cal	=	4.184 J
1 Torr	=	1 mm Hg
760 torr	=	1 atm

Metric Abbreviations

meter	m
centimeter	cm
millimeter	mm
nanometer	nm
liter	L
milliliter	mL
kilogram	kg
gram	g
milligram	mg
mole	mol

Temperature Conversion Formulas

$$°C = \frac{(°F - 32)}{1.8}$$

$$°F = 1.8 \ °C + 32$$

$$K = °C + 273$$

APPENDIX 5

Solubility Table

	$C_2H_3O_2^-$	AsO_4^{3-}	Br^-	CO_3^{2-}	Cl^-	CrO_4^{2-}	OH^-	I^-	NO_3^-	$C_2O_4^{2-}$	O^{2-}	PO_4^{3-}	SO_4^{2-}	S^{2-}	SO_3^{2-}
Al^{3+}	*aq*	I	*aq*	–	*aq*	–	I	*aq*	*aq*	–	I	I	*aq*	d	–
NH_4^+	*aq*	*aq*	*aq*	*aq*	*aq*	*aq*	*aq*	*aq*	*aq*	*aq*	–	*aq*	*aq*	*aq*	*aq*
Ba^{2+}	*aq*	I	*aq*	I	*aq*	I	sl. *aq*	*aq*	*aq*	I	sl. *aq*	I	I	d	I
Bi^{3+}	–	sl. *aq*	d	I	d	–	I	I	d	I	I	sl. *aq*	d	I	–
Ca^{2+}	*aq*	I	*aq*	I	*aq*	*aq*	I	*aq*	*aq*	I	I	I	I	d	I
Co^{2+}	*aq*	I	*aq*	I	*aq*	I	I	*aq*	*aq*	I	I	I	*aq*	I	I
Cu^{2+}	*aq*	I	*aq*	I	*aq*	I	I	–	*aq*	I	I	I	*aq*	I	–
Fe^{2+}	*aq*	I	*aq*	sl. *aq*	*aq*	–	I	*aq*	*aq*	I	I	I	*aq*	I	sl. *aq*
Fe^{3+}	I	I	*aq*	I	*aq*	I	I	–	*aq*	*aq*	I	I	*aq*	I	–
Pb^{2+}	*aq*	I	I	I	I	I	I	I	*aq*	I	I	I	I	I	I
Mg^{2+}	*aq*	d	*aq*	I	*aq*	*aq*	I	*aq*	*aq*	I	I	I	*aq*	d	sl. *aq*
Hg_2^{2+}	sl. *aq*	I	I	I	I	sl. *aq*	–	I	*aq*	I	I	I	I	I	–
Hg^{2+}	*aq*	I	I	I	*aq*	sl. *aq*	I	I	*aq*	I	I	I	d	I	–
K^+	*aq*	*aq*	*aq*	*aq*	*aq*	*aq*	*aq*	*aq*	*aq*	*aq*	*aq*	*aq*	*aq*	*aq*	*aq*
Ag^+	sl. *aq*	I	I	I	I	I	–	I	*aq*	I	I	I	I	I	I
Na^+	*aq*	*aq*	*aq*	*aq*	*aq*	*aq*	*aq*	*aq*	*aq*	*aq*	*aq*	*aq*	*aq*	*aq*	*aq*
Zn^{2+}	*aq*	I	*aq*	I	*aq*	I	I	*aq*	*aq*	I	I	I	*aq*	I	I

Key: *aq* = Soluble in water   I = Insoluble in water (less than 1 g/100 g H_2O)
   sl. *aq* = Slightly soluble in water   d = Decomposes in water

APPENDIX 6

Vapor Pressure of Water

Temperature (°C)	Vapor Pressure torr (or mm Hg)	Temperature (°C)	Vapor Pressure torr (or mm Hg)
0	4.6	26	25.2
5	6.5	27	26.7
10	9.2	28	28.3
15	12.8	29	30.0
16	13.6	30	31.8
17	14.5	40	55.3
18	15.5	50	92.5
19	16.5	60	149.4
20	17.5	70	233.7
21	18.6	80	355.1
22	19.8	90	525.8
23	21.2	100	760.0
24	22.4	110	1074.6
25	23.8		

APPENDIX 7

Boiling Points of Liquids

Liquid	Boiling Point °C
Acetone	56.5
Ethanol	78.4
Diethyl ether	34.6
Methanol	64.7
1-propanol	82.5
Water	100.0

APPENDIX 8

Waste Disposal Requirements for Each Experiment

Listed below are special waste containers specified in the experiments for student disposal of waste. Where students are instructed to dispose of wastes in the sink, or where the experiment does not generate waste, the requirements are listed as NONE.

We use the same Waste Heavy Metal bottle for many experiments by combining all the ions poured into it on the label. The same can be done for Organic Solvent Waste bottles.

Exp	Title	Waste Containers That Should Be Available to Students	
1	Laboratory Techniques	Waste Heavy Metals (Pb^+) Waste PbI_2 on filter paper	bottle jar
2	Measurements	Container for waste glass	
3	Prep. and Prop. of Oxygen	Recycled 9% H_2O_2, unreacted Büchner funnel-vacuum flask setup for disposal of MnO_2	bottle jar
4	Prep. and Prop. of Hydrogen	Recycled Mossy Zinc, rinsed Unreacted metal strips	jar jar
5	Calorimetery and Specific Heat	None	
6	Freezing Points	Waste Acetic/Benzoic Acid Mixture	bottle
7	Water in Hydrates	Waste Heavy Metal Residues (Cu^{2+}, Zn^{2+}, Sr^{2+}, Ba^{2+})	jar
8	Water, Solutions and pH.	Waste Organic Solvents (decane)	bottle
9	Properties of Solutions	Waste Organic Solvent (decane) Waste Kerosene Mixtures Waste Heavy Metal Solutions (Ba^{2+})	bottle bottle bottle
10	Composition of Potassium Chlorate	Waste Heavy Metals (Ag^+) Unused $KClO_3$	bottle bottle
11	Double Displacement Reactions	Waste Heavy Metals (Ag^+, Ba^{2+}, Cu^{2+}, Zn^{2+})	bottle

Exp	Title	Waste Containers That Should Be Available to Students	
12	Single Displacement Reactions	Waste Heavy Metals (Ag^+, Cu^{2+}, Pb^{2+}, Zn^{2+})	bottle
13	Ionization—Electrolytes and pH	None (Students do not handle the heavy metal solutions in the demonstration.)	
14	Identification of Selected Anions	Waste Organic Solvents (decane)	bottle
		Waste Heavy Metals (Ag^+, Ba^{2+})	bottle
15	Quantitative Preparation of KCl	None	
16	EM Energy and Spectroscopy	Waste Heavy Metals (Ni^{2+}, MnO_4^-)	bottle
17	Lewis Structures/Molecular Models	None	
18	Boyle's Law	None	
19	Charles' Law	None	
20	Liquids—Vapor Pressure and Boiling Points	None	
21	Molar Volume of a Gas	Büchner funnel-vacuum flask-setup for disposal of MnO_2	
22	Neutralization—Titration I	None	
23	Neutralization—Titration II	None	
24	Chemical Equilibrium	Waste Heavy Metals (Ag^+, Co^{2+}, Cu^{2+})	bottle
25	Heat of Reaction	None	
26	Distillation of Volatile Liquids	Recycled Ethanol/Ethanol Distillate	bottle
27	Hydrocarbons	Waste Organic Solvents	bottle
		Waste Heavy Metals (Ag)	bottle
28	Alcohols, Esters, Aldehydes, Ketones	Waste Organic Solvents	bottle
		Waste Heavy Metals (MnO_4^-, Ag^+)	bottle

Periodic Table of the Elements

Atomic masses are based on carbon-12. Elements marked with † have no stable isotopes. The atomic mass given is that of the isotope with the longest known half-life.

Key: Atomic number → Symbol → Name → Atomic mass (example: 11, Na, Sodium, 22.99)

Current ACS and IUPAC — Preferred U.S.

Transition Elements · **Inner Transition Elements**

Group	1 IA	2 IIA	3 IIIB	4 IVB	5 VB	6 VIB	7 VIIB	8 VIII	9 VIII	10 VIII	11 IB	12 IIB	13 IIIA	14 IVA	15 VA	16 VIA	17 VIIA	18 0 (Noble Gases)
1	1 H Hydrogen 1.008																	2 He Helium 4.003
2	3 Li Lithium 6.941	4 Be Beryllium 9.012											5 B Boron 10.81	6 C Carbon 12.01	7 N Nitrogen 14.01	8 O Oxygen 16.00	9 F Fluorine 19.00	10 Ne Neon 20.18
3	11 Na Sodium 22.99	12 Mg Magnesium 24.31											13 Al Aluminum 26.98	14 Si Silicon 28.09	15 P Phosphorus 30.97	16 S Sulfur 32.07	17 Cl Chlorine 35.45	18 Ar Argon 39.95
4	19 K Potassium 39.10	20 Ca Calcium 40.08	21 Sc Scandium 44.96	22 Ti Titanium 47.87	23 V Vanadium 50.94	24 Cr Chromium 52.00	25 Mn Manganese 54.94	26 Fe Iron 55.85	27 Co Cobalt 58.93	28 Ni Nickel 58.69	29 Cu Copper 63.55	30 Zn Zinc 65.39	31 Ga Gallium 69.72	32 Ge Germanium 72.61	33 As Arsenic 74.92	34 Se Selenium 78.96	35 Br Bromine 79.90	36 Kr Krypton 83.80
5	37 Rb Rubidium 85.47	38 Sr Strontium 87.62	39 Y Yttrium 88.91	40 Zr Zirconium 91.22	41 Nb Niobium 92.91	42 Mo Molybdenum 95.94	43 Tc Technetium 97.91†	44 Ru Ruthenium 101.1	45 Rh Rhodium 102.9	46 Pd Palladium 106.4	47 Ag Silver 107.9	48 Cd Cadmium 112.4	49 In Indium 114.8	50 Sn Tin 118.7	51 Sb Antimony 121.8	52 Te Tellurium 127.6	53 I Iodine 126.9	54 Xe Xenon 131.3
6	55 Cs Cesium 132.9	56 Ba Barium 137.3	57 La* Lanthanum 138.9	72 Hf Hafnium 178.5	73 Ta Tantalum 180.9	74 W Tungsten 183.8	75 Re Rhenium 186.2	76 Os Osmium 190.2	77 Ir Iridium 192.2	78 Pt Platinum 195.1	79 Au Gold 197.0	80 Hg Mercury 200.6	81 Tl Thallium 204.4	82 Pb Lead 207.2	83 Bi Bismuth 209.0	84 Po Polonium 209.0†	85 At Astatine 210.0†	86 Rn Radon 222.0
7	87 Fr Francium 223.0†	88 Ra Radium 226.0†	89 Ac** Actinium 227.0†	104 Rf Rutherfordium 261.1†	105 Db Dubnium –	106 Sg Seaborgium –	107 Bh Bohrium –	108 Hs Hassium –	109 Mt Meitnerium –	110 Ds Darmstadtium –	111 Rg Roentgenium –							

Lanthanide Series 6 *

58 Ce Cerium 140.1	59 Pr Praseodymium 140.9	60 Nd Neodymium 144.2	61 Pm Promethium 144.9†	62 Sm Samarium 150.4	63 Eu Europium 152.0	64 Gd Gadolinium 157.3	65 Tb Terbium 158.9	66 Dy Dysprosium 162.5	67 Ho Holmium 164.9	68 Er Erbium 167.3	69 Tm Thulium 168.9	70 Yb Ytterbium 173.0	71 Lu Lutetium 175.0

Actinide Series 7 **

90 Th Thorium 232.0	91 Pa Protactinium 231.0	92 U Uranium 238.0	93 Np Neptunium 237.0†	94 Pu Plutonium 244.1†	95 Am Americium 243.1†	96 Cm Curium 247.1†	97 Bk Berkelium 247.1†	98 Cf Californium 251.1†	99 Es Einsteinium 252.1†	100 Fm Fermium 257.1†	101 Md Mendelevium 258.1†	102 No Nobelium 259.1†	103 Lr Lawrencium 262.1†

Legend: ☐ Metals · ▨ Metalloids · ☐ Nonmetals

Atomic Masses of the Elements
Based on the IUPAC Table of Atomic Masses

Name	Symbol	Atomic Number	Atomic Mass	Name	Symbol	Atomic Number	Atomic Mass
Actinium*	Ac	89	227.0277	Mendelevium*	Md	101	258.0984
Aluminum	Al	13	26.981538	Mercury	Hg	80	200.59
Americium*	Am	95	243.0614	Molybdenum	Mo	42	95.94
Antimony	Sb	51	121.760	Neodymium	Nd	60	144.24
Argon	Ar	18	39.948	Neon	Ne	10	20.1797
Arsenic	As	33	74.92160	Neptunium*	Np	93	237.0482
Astatine*	At	85	209.9871	Nickel	Ni	28	58.6934
Barium	Ba	56	137.327	Niobium	Nb	41	92.90638
Berkelium*	Bk	97	247.0703	Nitrogen	N	7	14.00674
Beryllium	Be	4	9.012182	Nobelium*	No	102	259.1011
Bismuth	Bi	83	208.98038	Osmium	Os	76	190.23
Bohrium*	Bh	107	—	Oxygen	O	8	15.9994
Boron	B	5	10.811	Palladium	Pd	46	106.42
Bromine	Br	35	79.904	Phosphorus	P	15	30.973762
Cadmium	Cd	48	112.411	Platinum	Pt	78	195.078
Calcium	Ca	20	40.078	Plutonium*	Pu	94	244.0642
Californium*	Cf	98	251.0796	Polonium*	Po	84	208.9824
Carbon	C	6	12.0107	Potassium	K	19	39.0983
Cerium	Ce	58	140.116	Praseodymium	Pr	59	140.90765
Cesium	Cs	55	132.90545	Promethium*	Pm	61	144.9127
Chlorine	Cl	17	35.4527	Protactinium	Pa	91	231.03588
Chromium	Cr	24	51.9961	Radium*	Ra	88	226.0245
Cobalt	Co	27	58.933200	Radon*	Rn	86	222.0176
Copper	Cu	29	63.546	Rhenium	Re	75	186.207
Curium*	Cm	96	247.0703	Rhodium	Rh	45	102.90550
Darmstadtium*	Ds	110	—	Roentgenium	Rg	111	—
Dubnium*	Db	105	—	Rubidium	Rb	37	85.4678
Dysprosium	Dy	66	162.50	Ruthenium	Ru	44	101.07
Einsteinium*	Es	99	252.0830	Rutherfordium	Rf	104	261.1089
Erbium	Er	68	167.26	Samarium	Sm	62	150.36
Europium	Eu	63	151.964	Scandium	Sc	21	44.955910
Fermium*	Fm	100	257.0951	Seaborgium	Sg	106	—
Fluorine	F	9	18.9984032	Selenium	Se	34	78.96
Francium*	Fr	87	233.0197	Silicon	Si	14	28.0855
Gadolinium	Gd	64	157.25	Silver	Ag	47	107.8682
Gallium	Ga	31	69.723	Sodium	Na	11	22.989770
Germanium	Ge	32	72.61	Strontium	Sr	38	87.62
Gold	Au	79	196.96655	Sulfur	S	16	32.066
Hafnium	Hf	72	178.49	Tantalum	Ta	73	180.9479
Hassium*	Hs	108	—	Technetium*	Tc	43	97.09072
Helium	He	2	4.002602	Tellurium	Te	52	127.60
Holmium	Ho	67	164.93032	Terbium	Tb	65	158.92534
Hydrogen	H	1	1.00794	Thallium	Tl	81	204.3833
Indium	In	49	114.818	Thorium*	Th	90	232.0381
Iodine	I	53	126.90447	Thulium	Tm	69	168.93421
Iridium	Ir	77	192.217	Tin	Sn	50	118.710
Iron	Fe	26	55.845	Titanium	Ti	22	47.867
Krypton	Kr	36	83.80	Tungsten	W	74	183.84
Lanthanum	La	57	138.9055	Uranium	U	92	238.0289
Lawrencium*	Lr	103	262.110	Vanadium	V	23	50.9415
Lead	Pb	82	207.2	Xenon	Xe	54	131.29
Lithium	Li	3	6.941	Ytterbium	Yb	70	173.04
Lutetium	Lu	71	174.967	Yttrium	Y	39	88.90585
Magnesium	Mg	12	24.3050	Zinc	Zn	30	65.39
Manganese	Mn	25	54.938049	Zirconium	Zr	40	91.224
Meitnerium	Mt	109	—				

*This element has no stable isotopes. The atomic mass given is that of the isotope with the longest known half-life.

NAMES, FORMULAS AND CHARGES OF COMMON IONS

	Positive Ions (Cations)		Negative Ions (Anions)	
1+	Ammonium	NH_4^+	Acetate	$C_2H_3O_2^-$
	Copper(I)	Cu^+	Bromate	BrO_3^-
	(Cuprous)		Bromide	Br^-
	Hydrogen	H^+	Chlorate	ClO_3^-
	Potassium	K^+	Chloride	Cl^-
	Silver	Ag^+	Chlorite	ClO_2^-
	Sodium	Na^+	Cyanide	CN^+
2+	Barium	Ba^{2+}	Fluoride	F^-
	Cadmium	Cd^{2+}	Hydride	H^-
	Calcium	Ca^{2+}	Hydrogen carbonate	HCO_3^-
	Cobalt(II)	Co^{2+}	(Bicarbonate)	
	Copper(II)	Cu^{2+}	Hydrogen sulfate	HSO_4^-
	(Cupric)		(Bisulfate)	
	Iron(II)	Fe^{2+}	Hydrogen sulfite	HSO_3^-
	(Ferrous)		(Bisulfite)	
	Lead(II)	Pb^{2+}	Hydroxide	OH^-
	Magnesium	Mg^{2+}	Hypochlorite	ClO^-
	Manganese(II)	Mn^{2+}	Iodate	IO_3^-
	Mercury(II)	Hg^{2+}	Iodide	I^-
	(Mercuric)		Nitrate	NO_3^-
	Nickel(II)	Ni^{2+}	Nitrite	NO_2^-
	Tin(II)	Sn^{2+}	Perchlorate	ClO_4^-
	(Stannous)		Permanganate	MnO_4^-
	Zinc	Zn^{2+}	Thiocyanate	SCN^-
3+	Aluminum	Al^{3+}	Carbonate	CO_3^{2-}
	Antimony(III)	Sb^{3+}	Chromate	CrO_4^{2-}
	Arsenic(III)	As^{3+}	Dichromate	$Cr_2O_7^{2-}$
	Bismuth(III)	Bi^{3+}	Oxalate	$C_2O_4^{2-}$
	Chromium(III)	Cr^{3+}	Oxide	O^{2-}
	Iron(III)	Fe^{3+}	Peroxide	O_2^{2-}
	(Ferric)		Silicate	SiO_3^{2-}
	Titanium(III)	Ti^{3+}	Sulfate	SO_4^{2-}
	(Titanous)		Sulfide	S^{2-}
4+	Manganese(IV)	Mn^{4+}	Sulfite	SO_3^{2-}
	Tin(IV)	Sn^{4+}	Arsenate	AsO_4^{3-}
	(Stannic)		Borate	BO_3^{3-}
	Titanium(IV)	Ti^{4+}	Phosphate	PO_4^{3-}
	(Titanic)		Phosphide	P^{3-}
5+	Antimony(V)	Sb^{5+}	Phosphite	PO_3^{3-}
	Arsenic(V)	As^{5+}		

The anion charge groupings are: **1−** (Acetate through Thiocyanate), **2−** (Carbonate through Sulfite), **3−** (Arsenate through Phosphite).

NOTES

NOTES

NOTES

NOTES

NOTES